MARINE INVERTEBRATES IN THE AQUARIUM

Richard F. Stratton

Quarterly

yearBOOKS,INC.
Dr. Herbert R. Axelrod,
Founder & Chairman

Dominique De Vito
Chief Editor

yearBOOKS are all photo composed, color separated and designed on Scitex equipment in Neptune, N.J. with the following staff:

DIGITAL PRE-PRESS
Patricia Northrup
Supervisor

Robert Onyrscuk
Jose Reyes

COMPUTER ART
Patti Escabi
Sandra Taylor Gale
Candida Moreira
Joanne Muzyka
Francine Shulman

ADVERTISING SALES
Nancy S. Rivadeneira
Advertising Sales Director
Cheryl J. Blyth
Advertising Account Manager
Amy Manning
Advertising Director
Sandy Cutillo
Advertising Coordinator

©yearBOOKS, Inc.
© by T.F.H. Publications, Inc.
**1 TFH Plaza
Neptune, N.J. 07753
Completely manufactured in
Neptune, N.J. USA**

Designed by Sandra Taylor Gale
Cover design by Sherise Buhagiar

Introduction

Proud and arrogant humankind is a self-centered species that has difficulty seeing things from an objective perspective. Our propensity to use such a term as "invertebrate" as a group name in which to lump animals is one example, for the fact is that over ninety-five percent of all animals are invertebrates. But, obsessed as we are with our own phylum of chordates, we have the temerity to dismiss the rest of the animal kingdom as "invertebrates." (And it should be pointed out that the animal kingdom is only one of five kingdoms of living things, in most modern systems of classification. Thus, vertebrates are an even smaller percentage of living things,

dropping well down below the one percent mark.)

Recently, fascination with living reef tanks has made marine aquarists well aware of many invertebrates. However, the emphasis in books and articles about such systems has been on the corals, for the corals are the reef builders, and the reef is the home for a great number of tropical animals, vertebrate and non-vertebrate alike. This book is an attempt to turn the spotlight from the soft and stony corals to some of the other invertebrates. Some of them will complement the corals in a living reef tank, while others would decimate the corals—but we'll be sure to let you

Contents

What are Quarterlies?

Because keeping reef tanks is growing at a rapid pace, information on new equipment and new varieties is vitally needed in the marketplace. Books, the usual way information of this sort is transmitted, can be too slow. Sometimes by the time a book is written and published, the material contained therein is a year or two old…and no new material has been added during that time. Only a book in a magazine form can bring breaking stories and current information. A magazine is streamlined in production, so we have adopted certain magazine publishing techniques in the creation of this yearBOOK. Magazines also can be much cheaper than books because they are supported by advertising. To combine these assets into a great publication, we are issuing this yearBOOK in both magazine and book format at different prices.

Distributed in the UNITED STATES to the Pet Trade by T.F.H. Publications, Inc., One T.F.H. Plaza, Neptune City, NJ 07753; on the Internet at www.tfh.com; in CANADA Rolf C. Hagen Inc., 3225 Sartelon St. Laurent-Montreal Quebec H4R 1E8; Pet Trade by H & L Pet Supplies Inc., 27 Kingston Crescent, Kitchener, Ontario N2B 2T6; in ENGLAND by T.F.H. Publications, PO Box 15, Waterlooville PO7 6BQ; in AUSTRALIA AND THE SOUTH PACIFIC by T.F.H. (Australia), Pty. Ltd., Box 149, Brookvale 2100 N.S.W., Australia; in NEW ZEALAND by Brooklands Aquarium Ltd. 5 McGiven Drive, New Plymouth, RD1 New Zealand; in SOUTH AFRICA, Rolf C. Hagen S.A. (PTY.) LTD. P.O. Box 201199, Durban North 4016, South Africa; in Japan by T.F.H. Publications, Japan—Jiro Tsuda, 10-12-3 Ohjidai, Sakura, Chiba 285, Japan. Published by T.F.H. Publications, Inc.

MANUFACTURED IN THE
UNITED STATES OF AMERICA
BY T.F.H. PUBLICATIONS, INC.

know about those particular ones! The point is that there are a lot of invertebrates that are unsuitable for a reef tank housing living corals, but they are nonetheless worth keeping.

The invertebrate life found in terrestrial locations consists primarily of insects, spiders and worms. In the sea, the invertebrates are greater not only in number of different types but also for the most part in size. When invertebrates came ashore, they had to compensate for the increased pull of gravity by a reduction in size. (Even so, there are giant butterflies and beetles in tropical areas, but, of course, these are the exception.) One of the reasons for the popularity of the mini-reef tanks is in no small measure the result of the fact that they can showcase the beautiful and bizarre world of the invertebrate animals.

At once beautiful and surreal, sea invertebrates dazzle with their beauty and fascinate with their utter otherworldliness. When

A brain coral weighing more than 100 pounds is hardly suitable for most mini-reef aquariums. Still, it is not impossible for them to be maintained as a reef invertebrate by a knowledgeable aquarist with enough space. Photo by Courtney Platt, Cayman Islands.

science fiction writers try to stress that extra-terrestrial intelligent life is not likely to be humanoid, or even primate-like, for that matter, they often resort to some invertebrate body plan, just as an attempt to emphasize how different life forms from another planet could be.

Although we tend to think of vertebrate animals as the more complex ones, there are surprising degrees of complexity among invertebrate life forms. The fact is, however, that everything else being equal, evolution favors simplicity. (As just one example, humans have far fewer bones in the skull and jaw than do species of fish.)

The ability to see the beauty in many invertebrates may be an acquired taste, as we are "hardwired" to view the completely bizarre with suspicion and fear. The truth is that the invertebrates tend to take the prizes in all directions. That is, the ones that are undeniably beautiful, such as certain starfish, anemones, nudibranchs and crustaceans, take top prize in the beauty category, while the ones that are undeniably ugly, such as some sea cucumbers, pretty much take the prize in

The Australian sea cucumber *Paracucumaria tricolor* photographed in the Great Barrier Reef. Photo by Roger Steene.

This wonderful invertebrate tank features *Actinodiscus* corallimorph anemones. Photo by MP&C Piednoir.

that category, too. Of course, ugliness, like beauty, is in the eye of the beholder, but even the ugly invertebrates draw attention and engender a certain reluctant fascination.

It is often recommended to potential marine aquarists that they start with a fish-only tank and later on take on the keeping of invertebrates. The fact is, however, that some of the very first marine tanks kept, the "marine gardens" of the last century, were actually composed solely of invertebrate life. This approach can be recommended here, for the real trick is not so much the keeping of either marine fishes or marine invertebrates. The trick is in keeping the two different types together. There are several problems in this. One is that fish tend to feed upon invertebrates, so fish species have to be selected carefully. This is not an entirely one-way proposition, however, as many invertebrates prey upon fish. One of the most spectacular examples is the very harmless-looking cuttlefish, which will snatch your prize fish out of mid water right in front of your eyes!

Marine invertebrates with their slower metabolic rates can often be kept more easily than fishes. A tank consisting solely of invertebrates will normally put less of a load on the filtration system because of the lower metabolic rates of most invertebrates. So one option of the neophyte marine hobbyist is to start out with invertebrates, but keep only them. That is, the hobbyist should not try to keep fish species with them. And, of

This is a Mediterranean tank with fishes and invertebrates coming only from the Mediterranean Sea. Photo by MP&C Piednoir.

One of Nature's most weird animals is the cuttlefish _Sepiella amanis_. Photo by A. Woodward.

filtration, in addition to protein skimming. Obviously, they also do well in the natural aquarium, in which living rock and living sand are utilized as the major biological filtration. I say obviously because it is invertebrate animals, in addition to bacteria, that provide the filtration for such systems. That is, tiny anemones on live rock help provide additional filtration there, and the living sand contains various crustaceans, worms and sea cucumbers that help keep the sand constantly turned and clear of detritus. In all the so-called natural aquarium systems, all the animals in the tank help provide the biological filtration to supplement the work of the various bacteria and algae.

The living reef tank features strong sun-like lighting and super-clean water in order to keep corals, _Tridacna_ clams and anemones. Like fishes, invertebrate animals do best in tanks with good-quality water, but only the corals, certain clams and certain anemones need high intensity wide-spectrum lighting. As a matter of fact, even some of the coral species don't need it.

Although we won't cover the corals in any detail in this book, there will be an attempt to highlight many invertebrate species that can do well in a reef tank even though they can also thrive without the exotic lighting; some of them can even do well in a tank full of fish. Once you have developed a little experience at keeping invertebrates and gotten to know them well, it can be surprising how versatile and creative you become at putting them on display.

course, he should make an attempt to keep some of the easier invertebrates to maintain. That helps to inculcate confidence as well as enthusiasm. When you have your "sea legs," so to speak, you can try keeping some selected species of fishes with your invertebrates, but for right now let's concentrate on keeping invertebrates alone.

Even if you have already kept a fish-only tank, a good approach is still to set up a separate tank of invertebrates and develop an understanding of their care before attempting to keep them with fish. It will be most difficult for fish fanciers to heed such advice, as many of them simply want invertebrates as decorations for the fish tank or to make it "seem more natural." However, a key factor in success with invertebrates is in developing a faculty of caring for them in a tank of their own and then, later on, adding carefully selected fish species.

Invertebrates can be kept in a variety of tanks, with various different filtration systems. Of course, the all-glass tank is the way to go, just as it is with keeping marine fishes. Although invertebrates are less adaptive to changes in pH and increases in metabolic byproducts, such as ammonia and nitrite, they also, because of their lower metabolic rates, produce less of those substances, and they are also less likely to contribute to a decline in the pH.

The point is that invertebrates will generally do well in tanks that are kept with undergravel filtration, as well as compound filtration that relies upon wet/dry filtration, mechanical filtration and chemical

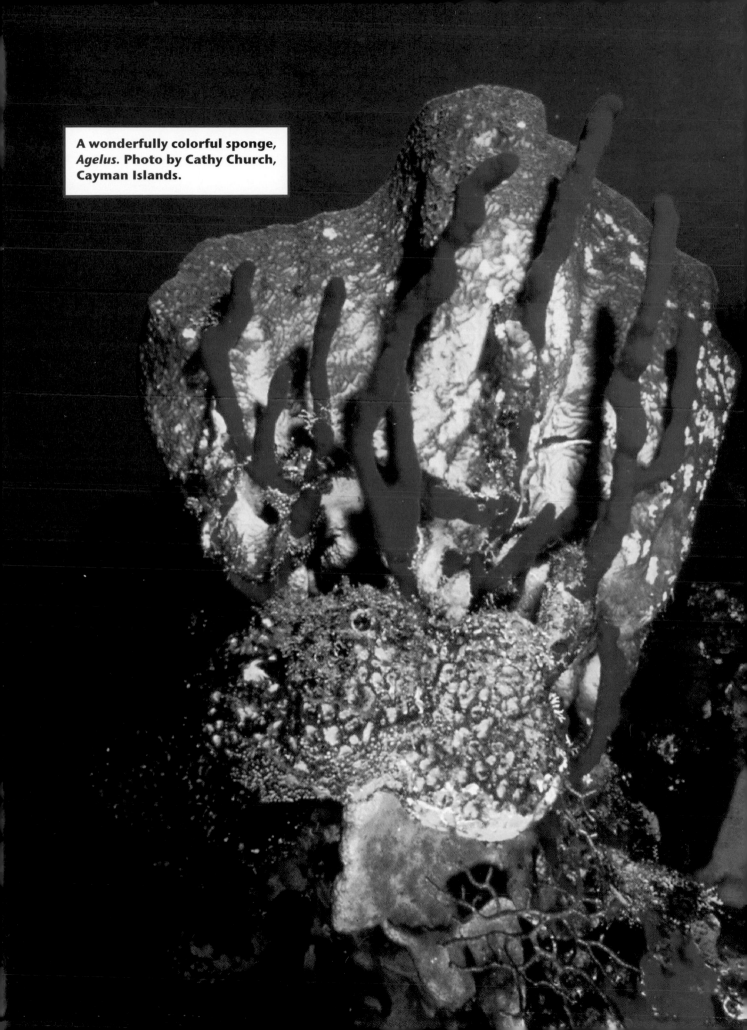

A wonderfully colorful sponge, *Agelus*. Photo by Cathy Church, Cayman Islands.

TYPES OF INVERTEBRATES

In this chapter we'll go over the different groups of invertebrates in phylogenetic order. That is, we will discuss them in the order that current thinking in biology has them appearing in the evolutionary scheme of things. In order to understand just how large these groups of invertebrates are, we need to understand a tiny bit about the classification system used by biologists. The point is that it has different levels of inclusion. For example, the kingdom Animalia (Latin or Greek is often used in science in order to have universal terms for these classifications) would include all animals. The class of birds (Aves) includes all bird species, just as Mammalia includes all mammals. As we move down the classification ladder, the genus includes only very similar species. For example, *Amphiprion* includes only clownfishes, and the fact is that one species of clownfish is even put in a different genus (*Premnas*). However, all clownfishes, together with other damsels, belong to the family Pomacentridae. The family places similar genera (plural of genus) under one grouping.

It is interesting to note that we humans are in our very own family all by ourselves. That is one advantage to being the species doing the classifying! In all fairness to ourselves, I should mention that whenever a lot of research is done on a

The clownfish *Amphiprion ocellaris* is always associated with an anemone in Nature. Photo by John Manzione.

particular group of organisms, a lot of "splitting" takes place. That means that the organisms are more often placed in separate species, genera and families. Also, we do have extinct (fossil) relatives in our family and in our genus. Incidentally, our family is Hominidae, our genus is *Homo*, and our species name is *sapiens*. *Homo* means humankind (or Man), and *sapiens* means "wise." Another advantage of our doing the naming is that we get to give ourselves a good name! Notice that all species have two names, the generic name and the species name. (When three names are in the mix, you're dealing

with a subspecies.) Also, notice that the species name is not capitalized, while the generic name is. Another point worth mentioning is that it is conventional to print the name in italics to denote the non-English language, in this case Latin. In the future, you may notice that newspapers and magazines not specializing in biology tend to misuse scientific names in various ways, ranging from failure to capitalize only the generic name to failure to utilize italics. (Isn't this fun? Now we can start finding fault with others and feel smug in the process!)

The different levels of

classification are the following:

Kingdom
 Phylum
  Class
   Order
    Family
     Genus
      Species

This *Premnas* was collected and photographed by Dr. Herbert R. Axelrod in Marau, British Solomon Islands.

Biology teachers teach their students to use a memory device, such as the sentence "Kings play chess on fancy gilded sets." Each word stands for a level of classification. One of my science classes contained a beautiful student named Clara, who drove the boys to such distraction that they came up with the following substitute sentence: "Kiss pretty Clara on face; get slapped!"

In any case, we humans belong to the kingdom Animalia, phylum Chordata, class Mammalia, order Primates, family Hominidae, genus *Homo*, and, finally, species *sapiens*.

If you are as old as I am, you remember a time that there were simply two Kingdoms of living things: Plantae and Animalia, plants and animals. With increasing knowledge in biology, the distinctions have been refined to include five kingdoms, sometimes six. The bone of contention here is whether viruses are living organisms. There are a lot of reasons not to consider viruses as living organisms, and yet they do mutate and they do reproduce, even if it is obligatory for them to have a host cell in order to do so. Most people who study viruses (virologists) look upon viruses as living and advocate a sixth kingdom for them. In the meantime, the most primitive kingdom is Prokarota (or Monera), which includes all the organisms without a nucleus for containing the genetic material. These are all bacteria and include blue-green algae, which biologists now mostly call cyanobacteria. The next kingdom is Protista, which includes single-celled organisms that enclose the genetic material in a nucleus. Then there are Plantae, Fungi and Animalia. The erection of the additional kingdoms has helped refine what is known about organisms in biology. A famous problem in the early days under the old system was whether single-celled organisms, such as *Euglena*, were in the plant kingdom or the animal kingdom, as it had the characteristics of both. That is, it could move about like an animal, but it had the power of photosynthesis, to use the energy of the sun to produce nutrients, just as plants do.

In our quick survey of

Sometimes biologists who are doing intensive work on a particular group of organisms will insert a subfamily or superfamily to further delineate organisms with similarities. For example, clownfishes are sometimes treated as a subfamily (Amphiprioninae), and cichlids, pomacentrids, surf perches and wrasses have been grouped into a superfamily (Labroidea). Similarly, there can be subclasses and subphyla.

Euglena, **having both plant and animal characteristics.**

invertebrates, we are going to discuss them at the phylum level. In other words, we will be looking at all of these non-vertebrates at the same level that we are when we are included with all the other vertebrates (because we belong to the phylum Chordata and the subphylum Vertebrata). At the phylum level, we are included with fishes, frogs, chickens and mountain lions. Anyway, you get the idea: generally speaking, these will be very large groupings of organisms. However, it is worth going over because it will help give us a scanty framework of how invertebrates are related. So let's get started.

PHYLUM PORIFERA

Sponges are the most primitive animals that an aquarium hobbyist will encounter. They are so distinct that sponge biologists sometimes advocate placing them in a kingdom of their own, the Parazoa. They represent an evolutionary dead end, having given rise to no other known phylum. Their gross structure is unique among animals: there is no mouth or true digestive cavity, no muscles, no nervous system. For that

matter, there are no organs at all in the usual sense of the word.

These animals are quite different in shape and color; in fact, as is the case with many corals, the shape of the animal is determined by currents and other habitat conditions. For that reason, it is difficult to classify sponges at a glance—or even a long look. Specialists identify them by microscopic examinations of their spicules and other structures.

Sponges are divided into four main groups according

A sponge of the genus *Agelus* **being investigated by a shrimp** *(Rhynchocinetes).* **Photo by Dr. Patrick L. Colin.**

to their type of skeleton. The Calcarea are the simplest sponges, and they have calcareous spicules. The Hexactinellida have six-rayed siliceous spicules and include the beautiful deepwater glass sponges. The Sclerospongia have massive lime skeletons composed of calcium

carbonate, spicules and organic fibers. These sponges are often mistaken for corals. The Demospongia is the largest group and encompasses species with variously shaped siliceous spicules, a spongin skeleton or a mixture of both. This group includes the bath sponges and the boring sponges. The bath sponges have been harvested for centuries for use in bathing. The name "boring sponges" arises from their habit of boring into rocks, stony corals, shells and ships. The demospongids are of various shapes and coloration and are the ones of most interest to aquarists.

Sponges are filter feeders, pumping the water in through inhalant pores, each called an ostium, and out through exhalant pores, each one called an osculum. (Often there is only one large osculum, however). There are various structures, such as flagella for directing the water, but sponges are so simple that it is possible to run them through a blender and have them live! Sponges are one of those animals in which the distinction is somewhat blurred as to whether it is a colony of different specialized cells or a distinct organism. Scientific opinion supports the latter view, but there is a thin line between sponges and certain colonies of protists living together in a cooperative community.

PHYLUM CNIDARIA

Cnidarians (also called coelenterates) vary in appearance and range in size from tiny hydra to coral colonies many miles across, as in the Great Barrier Reef off the northeast coast of Australia. They are all characterized by radial symmetry. This simply means that if you cut through a coelenterate, you will find the organs arranged in an even circle around a central axis. The body is basically a simple sack serving as a stomach, with a single opening that serves as both a mouth and as the exit for waste products. The opening is usually surrounded by tentacles armed with tiny stinging cells called "nematocysts" that are used for catching food.

The coelenterates consist of at least ten thousand known species and are divided into four classes:

HYDROZOA: This group contains the tiny hydroids and also the complex floating colonies that make up the infamous Portuguese man of war. The fire corals that cause painful "hot spots" to skindivers are also included in this group.

CUBOZOA: These are similar to jellyfish and include the Australian sea wasps, which are capable of quickly killing a person with their sting.

SCYPHOZOA: This group includes all the jellyfish, which are well named, as they are over ninety percent water. Most have an umbrella-shaped structure that propels them through the water, even though they always drift with the current. Most species are carnivorous and catch animals with their tentacles, which are armed with nematocysts. A few species are a type of filter feeder, and at least one lives off the

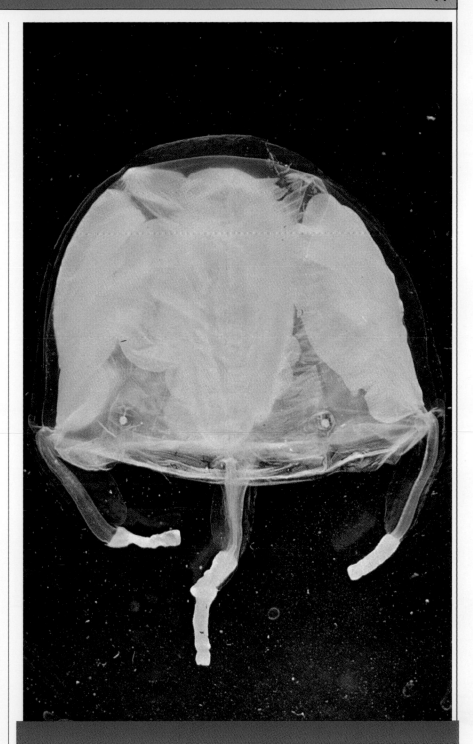

The sea wasp *Carybdea* has stung a lot of swimmers. Their stings can be neutralized utilizing Adolph's Meat Tenderizer. This product contains the papaya enzyme papinase, which can be used to neutralize most marine skin irritations, including Portuguese Men-of-War. Of course, ripe papaya can be rubbed on the irritation. Photo by P. Larson.

This colorful gorgonian is called *Corallium rubrum*. Photo by MP&C Piednoir.

commensal zooxanthellae in its body, so the only job of the host is to keep itself in a sunny location near the surface, where the zooxanthellae can perform photosynthesis and share nutrients with the "keeper."

ANTHOZOA: This is the largest group of coelenterates.

They are of the greatest interest to aquarists, because they include the sea anemones and the corals. Some species are strictly carnivorous, trapping prey with their stinging tentacles, while others rely upon zooxanthellae for part or all of their nutrition. Corals nearly

always live in colonies and can be soft or stony, with the stony corals generally being the reef builders.

Sea anemones have been described as upside-down jellyfish. (The same might be said of the tiny coral polyps, which look superficially like miniature sea anemones.) The base is a suckerlike disk that keeps the anemone securely in one spot, although they are capable of movement, much to the consternation of aquarists! Generally speaking, they stay put, but the ability to move can be triggered by being exposed to a predatory starfish or sea slug. Also, a sea anemone may move about the tank if it is unhappy with water conditions. They have many different color forms and are popular with marine aquarium hobbyists for that reason. Some species are strictly predatory, while others rely upon the zooxanthellae for at least supplementing their food.

The enhanced lighting of modern reef aquaria provided a big breakthrough in keeping some of the most desirable tropical anemones, which need the zooxanthellae in order to live. Before the advent of today's powerful wide-spectrum lighting, some hobbyists were able to keep the anemones by providing them with direct sunlight, usually from skylights directly above the tank. Some anemones are not totally reliant upon the zooxanthellae and simply turn white if there is not sufficient lighting, but they still prosper if proper nutrients are available. Other anemones, usually from deep

The burrowing anemone *Pachycerianthus forreyi*, from Monterey Bay, California. Photo by Ken Lucas.

The hydroid *Polyorchis*.
Photo by R. Larson.

water, don't have any zooxanthellae and rely entirely upon their tentacles for their nutritional needs, capturing plankton and small crustaceans.

PHYLUM PLATYHELMINTHES

This large group of flatworms includes the various parasitic flukes and tapeworms that, of course, are avoided by aquarists. However, there are some detritus-feeding and

Trematoda and Cestoda, while the free-living flatworms are in the class Turbellaria. There are at least four thousand species in the class; the most colorful and desirable aquarium inhabitants belong to the family Pseudocerotidae and come from the Indo-Pacific.

PHYLUM ANNELIDA

This large group of invertebrates contains the segmented worms, many of

cells. There are three main groups of annelids, the Oligochaeta, the Hirudinea and the Polychaeta.

OLIGOCHAETA: This class consists mainly of the familiar terrestrial earthworms but also includes some important marine species. The diversity and abundance of marine oligochaetes in estuaries and sheltered coasts is of fairly recent knowledge. Another example of this group is the well known tubificid worms, which are often sold live as food for tropical fish.

HIRUDINEA: This class contains the bloodsucking leeches, well known as external parasites. It is interesting to note that they have become important experimental animals in the study of the nervous system, and the future cure of certain types of paralysis may result from work with these "disgusting" creatures. There are only a few marine species.

POLYCHAETA: This class contains the bane of marine aquarists, for the so-called bristleworms, which have been known to eat corals and even attack large anemones, are in this group. It must be admitted that some species of bristleworms are quite pretty, but they are capable of delivering a sting (from the bristles) even to humans, and they prey upon the other invertebrates. They could be kept in a tank of their own by any hobbyist sufficiently interested in them.

There are many other polychaets, but those of main interest to the marine aquarist are the tubeworms. In tubeworms the tentacles usually form a crown that catches food particles in the

One of the Caribbean jellyfishes. These are favorite foods of the large sea turtles. Photo by Cathy Church, Cayman Islands.

carnivorous flatworms that have a pleasing color to them.

Flatworms are the most primitive worms, and their ancestors occupied a key position on the evolutionary tree leading to higher animals. The flatworm gut is still a simple, blind-ended tube, and respiration occurs throughout the body surface, but there is an excretory system, and there is a primitive nervous system with a tiny brain.

The flukes and tapeworms belong to the classes

which are marine, but the most familiar is the common earthworm. The name comes from the Latin *anulus*, which means "ring." These animals have no solid skeleton, but they gain rigidity from hydraulic pressure in the fluid-filled body cavity. Annelids have a straight gut, with a mouth at one end and an anus at the other. Other examples of advancement are the good circulatory system and the presence in each segment of excretory organs and a compact mass of nerve

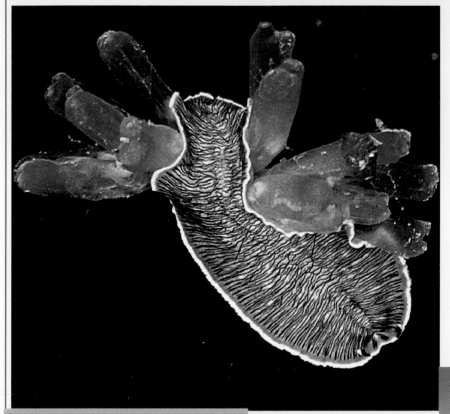

anatomy reveals numerous points of similarity to justify placing them all in the same phylum. Each possesses a foot, a muscular organ that, in the primitive state, serves both in locomotion and as a sturdy underpinning for the viscera. The foot of squids, octopuses and other cephalopods is modified into tentacles borne upon a head that is also equipped with eyes.

Mollusks also possess a mantle, a sheet of specialized tissue that covers most of the viscera like a body wall. The mantle secretes a shell, which provides an external armor in most molluscan groups but has been modified into an

Above: A marine flatworm, *Pseudoceros crozeri,* **eating an ascidian. Photo by Charles Arneson.**

Below: A tubeworm, *Sabellastarte indica,* **from Australia. Photo by Walt Deas.**

water. The particles are carried to the mouth by cilia on the tentacles. In small tubeworms the body surface suffices for respiration, but the larger ones need gills. In species with gills, the gills are usually situated among the tentacles and water is drawn past them by special movements of the body. The very popular Christmas tree worms are in this group.

PHYLUM MOLLUSCA

It is hard to imagine animals less alike in outward appearance than a snail, a clam and a squid. Yet each represents a class within the phylum Mollusca. Only a close comparison of their embryology and adult

spiders, mites, centipedes and insects, to name a few. Unfortunately, the trilobites became extinct during the Paleozoic Era, but the marine hobbyist can still keep horseshoe crabs, which are nearly as primitive! The original arthropods are believed to have evolved from annelid ancestors in Precambrian times. They acquired a unique armor, the exoskeleton or cuticle, which is composed of layers of protein and a strong flexible modified polysaccharide called chitin.

internalized support in slugs and squids and lost altogether in octopuses. Within the mantle cavity are found a small number of uniquely constructed feather-shaped gills. A scraping organ, the radula, is usually situated in the floor of the mouth.

There are many molluscan classes, as might be expected from such a diverse group, which includes everything from chitons (Amphineura) to bivalves (Bivalvia, or Pelecypoda) to nudibranchs (Gastropoda) to octopuses (Cephalopoda).

PHYLUM ARTHROPODA

In terms of numbers of species and diversity, the arthropods constitute one of the two or three most successful phyla on Earth. The phylum includes trilobites, crustaceans,

A marine snail, *Amplustre amalustrum*, is not so different looking when its mantle is expanded. Photo by Scott Johnson.

Presumably the exoskeleton evolved to provide protection, but, once evolved, it functions, much as our internal skeletons, as rigid components of the body to which organs and muscles can attach. In fact, it was in this group that muscles began to specialize to move the dissimilar body parts (as opposed to the annelid worms, in which the body parts are nearly all duplicates).

Right: Marine clams also are mollusks. This is *Tagellus plebius*, a favorite food within its range. Photo by Jim Bateman.

Barnacles are marine crustaceans famous for the strong glue which adheres them to the surface of their choice. Dental scientists tried to imitate this glue for repairing human dental caries, but they were unsuccessful in synthesizing the material. Photo by U. Erich Friese.

The crustaceans (class Crustacea) are the dominant arthropods of the sea. One group alone, the algae-feeding copepods, are so numerous in plankton communities that they may well be the most abundant of all groups of animals. Although crustaceans are distinctive in many ways, they can be characterized most simply as arthropods with two pairs of antennae. The most familiar crustaceans include shrimps, sowbugs, sand fleas, lobsters, crayfish and crabs. There is also a vast array of forms similar to presumed ancestral arthropods, and many bear a superficial resemblance to shrimps. The barnacles (subclass Cirripedia) have developed a wholly sedentary existence. The unique calcareous shells of these animals cause them to resemble mollusks, but, as the great zoologist Louis Aggasiz remarked over a century ago, a barnacle is "nothing more than a little shrimp-like animal, standing on its head in a limestone house and kicking food into its mouth."

PHYLUM ECHINODERMATA

This phylum includes starfish, sea urchins, sea cucumbers and crinoids, to name a few. Echinoderms are truly unusual, so distinctive in body plan from other animals that, if they were not so familiar as sea animals, they might seem to have originated on some other world. Their single most characteristic feature is the water-vascular system, an array of canals and tubelike appendages that serve several functions simultaneously, the principal ones of which are locomotion and the capture of food.

Although the total functioning of the water-vascular system is not completely understood, the

Goose barnacles, *Lepas anatife*. Photo by MP&C Piednoir.

mechanical operation of the basic unit, the tube foot, is clear enough. Each tube foot is a little adhesive organ that creates suction by hydraulic expansion and contraction. On touching a surface, the terminal sucker on the end of the tube foot draws back slightly, creating a small vacuum. Adhesiveness is simultaneously increased by the secretion of a sticky substance around the sucker. With hundreds of tube feet acting simultaneously, a starfish can exert enormous force. It can grasp a clam in its arms, anchor the arms with tube feet and, by steady contraction of muscles in the arms, gradually pull the shell apart.

Echinoderms are distinctively armored. They possess an internal skeleton of calcareous plates that are either articulated, permitting flexibility, or fused to form a rigid skeletal box. Warty or spiny projections extend outward as extra protective devices. The word echinoderm means "spiny skin." Most members of the phylum are radially symmetrical, but embryological studies show the condition to have been derived secondarily from the bilateral symmetry of ancestral forms.

The echinoderms are a very

The red starfish *Asterias forbesi* is a member of the phyllum Echinodermata. Photo by Jim Bateman.

An interesting starfish suitable for the reef aquarium, *Fromia monilis*. Photo by Dr. Herbert R. Axelrod.

ancient and diverse group. In addition to the starfish (class Asteroidea), there are four other major living groups. Brittle stars (class Ophiuroidea) resemble starfish but have long, flexible arms that allow them to move rapidly over the bottom or even, in some species, to swim. The arms tend to break off when grasped; hence the name "brittle starfish." (Many scientists encourage the use of the popular name "sea stars," since starfish are not fish; however, the public continues to call them by the name to which it is accustomed. The name originated from an ancient system of classifying all living things by the medium in which they lived. All things that flew were birds and all animals in the ocean were fish. Thus we have jellyfish and starfish, well knowing that they are not fish.)

Echinoderms have a number of technical features allying them to vertebrates, and that is one reason that this ancient group is placed so far along on the list.

PHYLUM CHORDATA

All right. We were only going to include invertebrates in this quick survey, so why do we have chordates? That is because, according to taxonomists, chordates don't have to have a notochord (a stiff supporting structure) throughout their life cycle. Tunicates, or sea squirts, have such a notochord only during their larval stage. In fact, the planktonic larvae look superficially like tadpoles. After that stage, they become sessile (fixed in place, like a plant) and no longer have the structure that places them in this phylum.

Sea squirts look very much like a sponge, but they are structurally much more complex. They also act much the same way as sponges act, even though they have muscles and organs. They draw in water and screen nutrients out of it. What were once gill slits become highly developed into a large basket used to filter food from the water. They are distinct from the sponges in appearance in

The famous blue starfish photographed in the Fiji Islands. Its scientific name is *Linckia laevigata*. Photo by Dr. Herbert R. Axelrod.

Left: The bat star *Patiria mimata*, shown with sea anemones. Photo by Dr. Herbert R. Axelrod

that they have one intake siphon and a similar-looking exhalant opening. The water is rapidly pumped through the animal, but the food is strained out and passes through a gut, unlike in sponges.

Although normally of little color, some sea squirts are colorful, also much like sponges. Small colonial specimens are often on live rock, but it will take a supply of plankton-type food to keep

Below: The crinoid *Himerometra*. Photo by Dr. Herbert R. Axelrod.

them alive. Large individual specimens are occasionally offered for sale in aquarium shops, but they need special attention in order to ensure their survival.

COMMENTS

This has been a quick overview of the invertebrate phyla. It would have been much easier to cover the vertebrates! It may seem overwhelming at first, but a hobbyist's assignment of animals to their correct phyla becomes second nature after a time of working with and keeping them. In the meantime, let's look at a few individual animals.

Right: Sea urchins belong to the class Echinoidea of the phylum Echinodermata; they are primarily vegetarian, grazing algae off rocks and other hard surfaces. The spines can cause wounds to bathers who step on them. Photo by U. Erich Friese.

POPULAR INVERTEBRATES

The idea in this chapter is to present some of the commonly kept marine invertebrates. We won't bother to arrange them phylogentically this time, but we'll figure out what they are and assign them to their phyla. It is interesting to note that not all of the popular invertebrates are easy to keep. Some of them are popular because of their appeal, but they have very special needs. Conscientious shop owners caution their customers about the difficult animals. That is fair to both the invertebrate and the customer (the vertebrate!), for the fact is that some customers enjoy a challenge and have the time and inclination to meet the special

Mixing hard and soft corals with other invertebrates and fishes can be an exciting sight, as you can see. But maintaining such an aquarium requires a great deal of skill and a knowledge of how these organisms live together.

A starfish, *Asterias*, eating a clam. Photo by Takemura and Suzuki.

needs of some of the more difficult specimens, while others will be discouraged in the hobby by unexpected failure with a given species. There was a time when "expect to fail" was standard advice applicable to any part of the marine hobby. Happily, such is no longer the case, and many of the popular invertebrates have gained their popularity from the fact that they are not only interesting to observe and fascinating in appearance but also easy to keep.

SOFT AND STONY CORALS

Of course, this is the group that we said that we would not emphasize, as a number of books feature information about keeping corals alive,

A rather barren mini-reef aquarium with *Centropyge loriculus* and *Gramma loreto* fishes with a few outstanding invertebrates. Most of the tank is overwhelmed by algal growth. Photo by R. Wederich.

and they form the basic information source for hobbyists who keep tanks that have corals as the primary inhabitants. The other occupants of the tank may be invertebrates, but they must be compatible with the corals, as the corals are the royalty in such tanks. (Or at least they are the animals that are emphasized, because the luxuriant growth of such colonies forms the basis for everything else that goes into the tank.)

Coral tanks are characterized by super-clean water and strong sunlike lighting of a very wide spectrum, but with an emphasis on the blue light. This calls for using exotic filtration, such as trickle filtration or/and fluidized bed filtration, as well as a protein skimmer. Recent years have seen the natural aquarium gain in popularity. The adherents of the natural aquarium tend to eschew the use of exotic filtration and rely instead upon the natural filtration that eventually results with the use of live rocks and living sand. That

means rocks that are taken from the ocean (or cultivated therein) and sand that is taken from the ocean (or, again, cultivated in the ocean or ocean water). The theory here is that the rocks and sand provide "homes" for the bacteria that will provide the biological filtration for the aquarium. There are different schools of thought on the establishment of the natural aquarium, and they are beyond the scope of this book; however, the most popular, and one of the most successful, schools is the

The sea squirt *Halocinthia papillosa*. Photo by MP&C Piednoir.

Berlin System, which advocates the use of a high-quality protein skimmer in addition to all of the live rocks.

Because of the zooxanthellae in their tissues, some of the corals never really need to be fed, but others need occasional feedings of planktonic-like food, while still others require feeding more than once a day. In any case, some of the most popular corals are the so-called meat or trumpet corals, lettuce corals and bubble corals. All of these corals are photosynthetic and don't need other than occasional supplementary feeding, and many hobbyists eschew even that. Some of the leather corals and other soft corals are not photosynthetic and therefore need to be fed. The point is, of course, that not all invertebrates are in need of high-intensity lighting, and it is our intention to delineate some of them. That is, we will cover them without restricting ourselves simply to those invertebrates that need photosynthetic (actinic) lighting or are compatible with such a setup; however, we will make note of whether or not they are compatible with corals and the bright lighting that they live in.

OCTOPUSES

Octopuses have been portrayed as ugly, evil creatures and true dangers to any human foolish enough to enter the ocean. I remember that as a small boy I had nightmares after reading the Sunday comics, in which Prince Valiant was thrown into a pit that contained an octopus. Later I read Victor Hugo's *Toilers of the Deep*, in which the octopus was depicted in the worst terms possible. In fact, Hugo, as I recall, depicted death from an octopus as being a thousand deaths, as each sucker leached the life from a person. Well, I still like Hal Foster (the cartoonist who did "Prince Valiant") and Victor Hugo, but they sure were wrong about the octopus.

The fact is that most octopus species are quite small. An exception is the U.S. West Coast species, *Octopus dofleini*, which gets

Dendronephthya, one of the soft corals of the order Alcyonacea, which also contains the "leather" corals. Photo by U. Erich Friese.

quite large. In fact, scuba divers have had contests of trying to drag such species off a rock or out of their "lair" without suffering anything other than humiliation when the octopus outmuscled the diver and escaped. Unfortunately, not enough specimens escaped, and the states of Oregon and Washington have had to

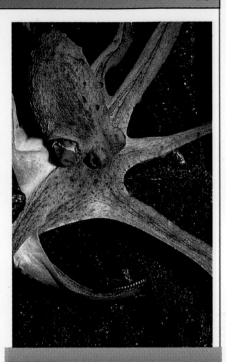

The octopus *Octopus vulgaris*. Photo by MP&C Piednoir.

establish state parks and provide legislation to protect these "monsters."

Of course, even small species have a mild venom, which is injected via the beak, but the venom's purpose is to immobilize a crab; it is rarely used on humans, even when they have the effrontery to capture specimens barehanded. The one species that is always warned about is the Indo-Pacific blue-ringed octopus, *Hapalochlaena maculosa*, a species not uncommonly offered as an aquarium resident even though its bite can be fatal to a human. The fact is that divers are not afraid of even this octopus, as it is quite shy, like all other octopuses, and the diver has to actively grasp the animal to get it to bite him.

An Octopus in the Aquarium
Let's get to the fun part:

keeping octopuses in the aquarium. Not only is it possible, but they can really be a lot of fun. There are lots of small species available that are offered in aquarium shops or can be ordered from your dealer. The trouble is that an octopus can cause a problem in the living reef tank if it is frightened and expels its "ink." This can harm delicate invertebrates, but the chances of the octopus's actually doing that are quite small. Since octopuses normally dine on crabs and shrimp, it would certainly be foolish to try to keep them in the same tank with these animals—unless you simply want your octopus well fed! Another problem with octopuses in the reef tank is

The Australian blue-ringed octopus, *Hapalochlaena maculosa*, found in southeastern Australia, has very variable blue markings. Photo by Keith Gillette.

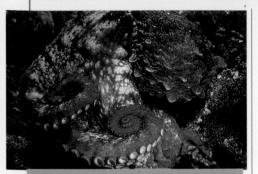

Octopus dofleini. **In the scheme of things, octopuses have been very successful in surviving through evolutionary history. Photo by Yuri F. Astafyev.**

could the octopus squeeze through a tight place to get out of the tank, it was even able to squeeze under a door to get out of the room!

An important point worth remembering here is that an octopus is a mollusk. That is, it belongs to the phylum Mollusca, which includes snails and slugs. We should therefore not be surprised that an octopus actively seeks a hiding place. In fact, if there is a shell in the tank, an

than many fishes. A recent study showed that octopuses kept under laboratory conditions could learn by example. That is, they could learn from watching other octopuses perform certain actions for food. Such ability is most definitely a sign of high intelligence. (To be perfectly honest, I should add that such studies were roundly criticized in peer review for a number of sound reasons. Even so, octopuses

that they can be hard on the water quality because of their high metabolic level. Having said that, though, I should mention that they certainly have been kept in reef tanks successfully. The point is that they can present special problems.

Although octopuses are generally regarded as very demanding in respect to water quality, several decades ago I kept a number of specimens that I had caught myself. The method of capture was simply to turn over rocks at low tide. Octopuses that had been stranded by the receding of the tide would occasionally be exposed when a rock was lifted up and were easily captured, even though they actively scurried away, looking much like a giant spider. I kept these specimens in tanks with just undergravel filters. In the process, I learned that if you don't provide a number of hiding places for an octopus, it will eventually get under your filter plate, as it is desperate for a dark retreat. I also learned that not only

One of the ocean's most interestingly colored animals is *Octopus horridus,* from the Indian Ocean. Photo by Alex Kerstich.

octopus will often show its mollusk affinity by not only getting into the shell but also by "wearing" it, much like a hermit crab, holding the shell in place with its tentacles.

One of the reasons that an octopus is a desirable aquarium resident, besides its novelty, is that it is a very intelligent animal; in fact, it may be the most intelligent of all the invertebrates. Certainly it is more intelligent

are certainly among the most intelligent of the invertebrates.)

Although shy by nature, octopuses become good pets and will take food from your fingers. They will become so tame that they will practically crawl into your lap! Having an octopus hug you is an interesting experience, and it is difficult to describe. The suction cups don't hurt, but the sensation of having them used on you is definitely a sticky one. Regarding food, you can give an octopus bits of fish or, preferably, bits of shrimp and crab. In nature, the octopus ambushes its prey by dropping over the top of it and holding it in place with the tentacles. And this can be seen in the aquarium if you have a crab or shrimp. Octopuses should not be fed more frequently than once a day, because they can pollute the tank with waste products. The hobbyist must be of steel will here, as once these animals have learned to associate their owner as a source of food, they can be quite persuasive in convincing him that he is underfeeding them.

Breeding Octopuses

Breeding octopuses in the aquarium is fraught with difficulties, because it is difficult to keep more than one species to a tank, for these are solitary animals in nature, as so many predators tend to be. Most species of octopus reach sexual maturity within a year. The mating ritual involves a male's inserting part of one arm, bearing sperm contained in membranous packets, into the female's genital opening. The specialized sexual portion of the male's arm detaches and remains with the female, and the male dies shortly thereafter. Not long after the mating, the eggs are deposited in a protected place by the female, who conscientiously guards them and keeps them clean and well aerated by circulating water over them with her siphon. Finally, she assists dozens to hundreds

(depending on the species) of her babies to be on their way as they hatch out. Carefully directed jets of water from the siphon tube disperse the young. Shortly after that, the female dies.

Although I don't know of any hobbyists who have bred an octopus species in the home aquarium, I have known several who happened to get females that laid eggs. Since fertilization is internal, the eggs eventually hatched. What is interesting here is that in nearly every case with which I am familiar, the female pried up the filter plate of the undergravel filter and used the underneath section as a "brooding cave" for laying her eggs and tending them. Only a few young were raised, and in some cases none at all were brought to maturity. The problem was trying to find foods that the tiny babies would eat. Newly hatched live brine shrimp and copepods were used in the successful cases.

The thrill of having baby octopuses in the aquarium in each instance was tempered by the death of the mother. Although this is a natural occurrence, it was difficult for the aquarist to accept. In every case, the intelligence and eventual tameness of the octopus had made it a favorite pet. It was, as one hobbyist told me, "like losing a member of the family."

How ironic that a member of such a feared and hated group should eventually be mourned like family! Well, anyway, the keeping of an octopus has much to recommend it; however, it should probably be kept by

itself, rather than in a community tank of invertebrates.

SEA APPLES

An echinoderm, the sea apple *Pseudocolochirus axiologus*, is one of the most popular invertebrates sold in aquarium shops. Unfortunately, this sea cucumber is a plankton feeder, so it needs very special attention. I just hope that all of the shop owners inform their customers.

As irresistible as these animals are, I would certainly not encourage their inclusion in a tank unless the owner is committed to providing a plankton-like food several times a day. Part of the appeal of the sea apple is that it has appendages that look absolutely otherworldly, as does the sea apple itself. Part of the weirdness involves the constant motion of its appendages as they scour the water for plankton to carry to the "mouth." Each takes a turn, but it is difficult to predict which appendage will go next. Of course, the echinoderms are very different-looking animals anyway, but echinoderms like the starfish are more familiar to us, so they don't seem quite so exotic. In any case, we will get back to some more practical echinoderms later.

THE ANEMONE TANK

I am thinking of tropical anemones for this tank, so the tank would need good filtration or a good protein skimmer to keep the water quality in excellent shape. The best choice would be one or more protein skimmers, as even biological filters (except

A sea apple, *Pseudocolochirus*. Photo by MP&C Piednoir.

Some filtration units combine a biological filter and a protein skimmer in one unit, thereby saving space. Photo courtesy of Cyclone Bak-Pak.

favorite spot and stay there. An important point here is that we are no longer restricted to the small anemones in a tank that is designed specifically for anemones. We can indulge ourselves with one of the giant anemones of the type that clownfish like to inhabit. If the tank is not of extraordinary size, however, we may need to restrict ourselves to just one specimen, as some of these anemones get big—really big! And they can be quite aggressive toward other anemones.

A different approach would be to keep a number of the smaller anemones. The

determining point is whether you want to emphasize a variety of anemones or have a spectacular display of one or two of the large anemone species. While a garden of anemones makes an interesting display, one huge anemone in the tank with an assortment of commensal animals, including clownfish and anemone crabs, is especially beautiful, and it never fails to attract attention. Even in these days of nature programs on television, it is surprising how many people have never heard of the symbiotic mutualism between an anemone and clownfish. Even if they have heard about it,

the ones with anaerobic bacteria, and they haven't been perfected yet) don't break down nitrates, and anemones are especially sensitive to nitrates.

In the reef tank, smaller anemones are selected, as large ones can be harmful to the corals. It is difficult to recognize it in animals as primitive as the coelenterates, but even they can be territorial. Thus coral colonies must be placed some distance from each other, as they will utilize specially adapted stinging cells to attack other corals infringing on their territory. The anemones are the same way, and the problem with them is that they can move around on their own. Fortunately, given the proper lighting and tank conditions, they tend to find a

The sea anemone *Condylactis gigantea*.

The magnificent anemone, *Heteractis magnifica*. Photo by Dr. Herbert R. Axelrod.

people are usually excited to actually see such a relationship on display.

It is possible to have an amalgamation of the two approaches (emphasis on several smaller anemones as opposed to emphasis on one or two large anemones) by keeping some hardy smaller anemones, such as the tube anemone *Pachycerianthus mana* and the Caribbean anemone *Condylactis gigantea*, along with one or two of the large *Heteractis* species. If you have a long enough tank, you could place one large anemone at one end and another at the other. It would be a good idea to have a large rock in the center of your tank; that will discourage either of the anemones from migrating toward the other one. Then you could have a colony of one species of clownfish at one end of the tank and another species at the other.

Even with just one large anemone and one species of clownfish, we are getting into a realm that is hard to beat in both appearance and interest. If anything rivals a reef tank, this would be it. Giant anemones, such as *Heteractis magnifica* (formerly identified as *Radianthus ritteri*) and *Stoichactis gigas*, the carpet anemone, are beautiful displays in themselves. But when you add the grace, movement and color of some clownfish, you truly have a breathtaking display.

It is difficult to say which species of clownfish is the most attractive. Use your own taste in picking. Be aware that clownfish are not quite as hardy as other damselfish. That is, they are more

The anemone fish *Amphiprion clarkii* taking refuge in the magnificent anemone, *Heteractis magnifica*. Photo by U. Erich Friese.

sensitive to shipping. So they should be acclimated separately from your anemones. A good approach is to get your anemones acclimated in your tank first. Once you are confident that they are used to the tank and doing well, you can pick out your clownfish to place with them.

It is extremely important that the fish be quarantined and free of disease before you place them into the tank. Not that they are going to give any diseases to the anemones, but fish are difficult to retrieve from such a tank, and you can't treat them for diseases in the tank with the anemones. For that reason, you will want to have a quarantine tank or to make sure that your dealer quarantines the fish for you (but expect to pay more for such fish). The quarantining and acclimating takes about a month, so it is no small undertaking on the part of the dealer to provide the

space and care for such a period of time.

Again, beauty is in the eye of the beholder as to which clownfish is the prettiest. One thing to take into consideration is the hardiness of the species. Although clownfish don't ship well as compared to other damsel species, they generally prosper in tanks once they have survived the shipping and become acclimated. A species that is especially hardy is the maroon clownfish, *Premnas biaculeatus*. This species does best when kept in pairs, and dealers often sell them as mated pairs. Surprisingly (when compared with other familiar fish species), the male is much smaller than the female. This species will accept many different anemones, and if there are no anemones in the tank, it will pester large coral polyps, trying to "establish a relationship" with them by rubbing against them. The clownfish often apparently succeed to the point that their body covering takes on enough of the polyp's coating that the polyp will stay extended rather than withdraw; however, repeated rubbing by the clownfish will damage or even kill the polyp. But I digress. The point is that this is a hardy species that can be kept as pairs and will thrive in a variety of large anemones.

A long-time favorite is the common clownfish, *Amphiprion ocellaris*. There is a variety of coloration in the species, but the best specimens are brilliant orange or red with white marks, edged in black, which look for all the world as though they had been painted on the fish. Like *Premnas*, this species also will accept many anemones, and it can be kept in groups of five or so on a single anemone. Captive-bred specimens are often available. They may cost more, but they are well worth the price, as they are more likely to be disease-free and to thrive in the aquarium.

Lest you think that you can keep only clownfish in your anemone tank, I should mention that it is possible to keep other fish, too, including juvenile angels. The main consideration here is whether the angels bother the anemones. Coral reef fish

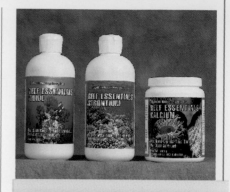

Manufacturers have made trace elements and other preparations for making the water in marine aquaria safer for fishes and invertebrates available in various forms. Photo courtesy of Boyd Enterprises.

Premnas biaculeatus, maroon clownfish, flirting with a sea anemone in which they wish to reside. Photo by Dr. Herbert R. Axelrod.

instinctively know enough to stay away from an anemone, but some angels, and other fish, too, will nibble at an anemone's tentacles. This should not be a serious problem in a tank that contains clownfish with the anemones, for the clownfish will protect the anemone from marauding fish.

One of the most beautiful displays would include some of the blue devils (numerous species of pomacentrids), as the bright blue coloration contrasts beautifully with the bright orange colors of the clownfish. Again, some blue damsels will pick at an anemone, but the clownfish will usually provide sufficient protection.

At this point I should apologize for talking about fish species in a book about invertebrate animals, but one of the secrets to a good invertebrate display is finding the proper fish to go with it, although certainly it must be admitted that many displays of invertebrates alone are enchanting.

In this era of emphasis on living reef systems, the anemone tank with its clownfish, crabs and other possible residents is often overlooked. More's the pity, because it is often the most beautiful of displays.

STARS IN OUR EYES

What is more suggestive of the sea than a starfish? Many people have never seen one alive, but everyone has seen the dried-out carapaces of various species sold around the world as novelties. It speaks highly of the success of all of the various starfish species that the sea has not

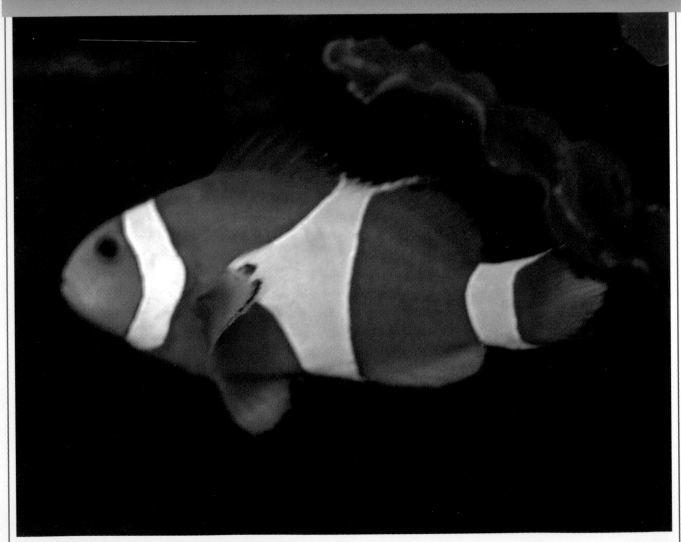

The anemone fish *Amphiprion ocellaris.* Photo by Dr. Karl Knaack.

been completely denuded of them, as almost every child has been given a starfish at least once, and they are a staple of all shell collections—even though they aren't shells. I well remember, many decades ago, a fellow wading out into the ocean at low tide and coming back with over twenty starfish stacked like pancakes. He would unload them in a basket and return for more. That was before any sort of era of conservation had begun, but I remember feeling somewhat resentful of the waste of all those animals just to decorate that guy's yard. (In retrospect, though, he was probably selling them as dried carapaces once they had cured.) Although there are still plenty of starfish in the ocean, it is no longer possible to collect them in the manner that was just mentioned.

There are still plenty of starfish for our tanks, though, but not all of them are equally suited. Some species, such as bat stars, are amazingly hardy, but other types of starfish are nearly impossible to keep alive. I have never seen such specimens offered in a dealer's tank, but the most common species found off the California coast is of that variety. They are a little large for a home tank anyway and are fierce predators on all manner of bivalves.

The fact is that nearly all starfish are a little suspect for the living reef tank, in the sense that even the ones that are basically scavengers may try to eat the polyps of stony corals. Let me list a few species that should work out in reef tanks. I have never known any of these species to harm corals, but an important point in selecting specimens is to get relatively small ones.

Ophiomastix venosa, a brittle starfish: Although the

Reef aquariums benefit from additional calcium that stabilizes alkalinity without affecting the pH, releases carbon dioxide to enhance photosynthesis and is biologically safe. Photo courtesy of Tropic Marin USA, Inc.

brittle starfish in general don't match our idea of what starfish "should" look like, this particular species has more personality than most of the others because it moves around more rapidly and quickly comes out of hiding when food is placed into the tank. It is of particular value in the reef tank because it is a scavenger and will help keep debris cleaned out of the many crevices invariably found in such a tank.

There are many species of brittle starfish. This one comes from the Caribbean, and it is frequently found in association with sea urchins, utilizing the protection of the spines. As might be supposed, the popular name comes from the propensity of such species to easily drop an arm if "trapped." Such arms normally regenerate; however, such "trauma" should be avoided, as too many incidents would most definitely take a toll on the animal. The animal can be handled if it is done gently and no attempt is made to grasp it by one of the rays, or "arms." It is fun to keep several of these types and watch them come bailing out of their hiding places when food is put into the tank.

Blue Starfish (*Linckia laevigata*): This species is undoubtedly one of the most dramatically colored of all the invertebrates, and the blue coloration makes a nice contrast in the reef tank to offset the many reds. To add to its desirability, this species is often available, and it is easy to keep. It feeds well and will scavenge, but many hobbyists simply place a bit of clam under the animal to speed up the process. This species won't sprint from one end of the tank to the other to get food as a brittle star will, but it is a beautiful decoration and an interesting addition to the tank. *L. laevigata* hails from the Indo-Pacific, as so many of the desirable species do, and in its natural habitat it is primarily a scavenger. So the blue starfish earns its keep not only by displaying during the daytime but also by helping to keep the tank picked up as well.

Red Starfish (*Fromia elegans*): This bright red

Yellow Starfish (*Fromia monilis*): If you want to keep starfish, you might as well have variety in coloration! This species varies from yellow to orange and usually has shades of each. It is very similar in habits and diet to the species already described. It is hardy and an excellent resident for the reef tank. It might be worthwhile at this point to mention that one of the great secrets to success with starfish is to obtain good specimens to begin with. Healthy specimens are eager

Although starfish are a special problem to keep in that we have to select the right ones and then make sure that we don't have animals that will hurt them, they are worth the effort. The biggest problem with them is that many people will ignore your exotic anemones and corals and will instead concentrate on your starfish. And with good reason, for starfish are the very essence of the sea. They are at once grotesque and beautiful, common and exotic. What more could we ask for? And all of the starfish mentioned are at home in the reef tank or in a tank with only dim lighting.

CUTTLEFISH

The first time I showed my sister-in-law a cuttlefish, she exulted, "Oh, what a cute animal! It looks just like a little old man with a beard!" The cuttlefish was simply hanging in the water, and I suppose it did look cute and cuddly, but I couldn't help smiling grimly at what she had said. Then I went on to explain to her that she wouldn't think the animal so cute if a fish or crab were thrown in there. I went on to explain that they were about as efficient and deadly as predators come.

I knew firsthand what a cuttlefish can do, as I had seen them at work, at both the original Sea World in San Diego and the Monterey Aquarium up in Steinbeck country. In each case, the cuttlefish was kept by itself, and I could see why when it was fed, for it seemed that the animal would eat just about anything that moved! In the

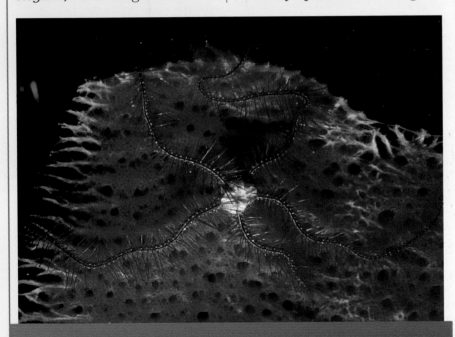

The brittle star *Ophiothrix svensenii*. Photo by Cathy Church.

species is not a threat to corals or other sessile invertebrates. It will happily dine upon bits of shrimp and shellfish. It should be kept in mind that all small starfish are endangered by predatory crabs or even large predatory starfish. So the emphasis is on small size for your starfish as well as for some of your other invertebrates, such as the crustaceans.

eaters and have a good tone. Limp bodies are not a good sign. Also, have the dealer look the specimen over for parasites. (Most knowledgeable and reputable dealers are going to do that anyway.) Starfish like good water quality, although some species tolerate subpar water better than others. The species mentioned are hardy but do require good water quality.

first locality, a fish was thrown in, and the cuttlefish pounced immediately. I am not at all sure that "pounced" is the word, as what really happened is that two specialized tentacles shot out and grabbed the fish with lightning-quick speed and jerked the hapless animal into its maw. One surprise was the distance that the two tentacles shot out to grab the fish, and the other was the speed at which they did so. The fish probably never knew what hit it, as the end was swift, and the beak and radula of the cuttlefish were as efficient as a blender. In the second locality, it was a large crab, and the result was the same.

Obviously, cuttlefish are not for your average marine hobbyist. For one thing, they are very demanding in regard to water quality. On the other hand, they can't be kept in most reef tanks, as they would quickly dispatch any fish or crustaceans the tanks contained. However, they are suitable for a tank in which there are only sessile invertebrates, which don't provide the movement that attracts the attention of the cuttlefish. Frankly, I would hesitate to place one in even a tank such as that, but I have seen them so displayed. The reason for my reservation is that reef tanks need really high-quality water, and the high metabolism of the cuttlefish puts a heavy strain on any filtration system. The best place for a cuttlefish is in a tank of its own with a good filtration system. I would suggest that the tank be equipped with a protein skimmer and a fluidized bed filter.

Feeding is obviously not a problem, as it will eat nearly anything alive that is placed into the tank. And I have talked with cuttlefish lovers who had trained their charges to eat thawed-out fish and crabs. The training simply consisted of dropping in live food and occasionally dropping in a thawed-out fish. Soon the quick-learning cuttlefish would take anything that hit the water, including pieces of crabmeat.

Products designed to re-move excess phosphates from the mini-reef aquarium are available at pet shops. Photo courtesy of Boyd Enterprises.

It is interesting that we see as many cuttlefish as we do among marine aquarists, although they are hardly common. The fact is that the animals are difficult to ship, and we have already discussed the other problems with them—the little habit they have of eating their tankmates! The driving force for keeping one of these animals is just that they are so fascinating in behavior. Like their cousins, the

octopuses, cuttlefish are highly intelligent and seem capable of learning to recognize people. In fact, they can look like a super-intelligent extraterrestrial being contemplating lesser organisms as they hover motionlessly in the tank. (At least, they can look that way to those of us who are highly imaginative—and science fiction fans, to boot!)

To the general public, the cuttlefish is mostly known by the cuttlebone that is taken from the captured Atlantic species *Sepia officinalis* and placed in bird cages to provide parrots and parakeets with calcium. Of course, the cuttlefish are not fishes at all; they are cephalopods, being

Marine Invertebrates
And Plants of the Living Reef

• How to identify most reef invertebrates from the Gulf of Mexico, Caribbean Sea, Florida and the Tropical Atlantic.
• For the Miniature Reef Aquarist • SCUBA Diver • Snorkeler • Ichthyologist
• Illustrated with 432 color photographs

Dr. Patrick L. Colin

close relatives to the octopus, giant squid and chambered nautilus. To place the cuttlefish in the modern scheme of classification, it is only necessary to remember that cuttlefish belong to the animal kingdom, the mollusk phylum, the cephalopod class, the order Dibranchia and the family Sepiidae and that most species of cuttlefish belong to the genus *Sepia*, which once provided the sepia-colored writing ink used for many years.

Now it is interesting to note a couple of points here. First of all, the mollusk phylum is one of the most successful of all the phyla, being second in number of species only to the arthropods, which includes crustaceans, spiders and insects. Mollusks are

primarily known for their shells; however, the octopus lacks a shell, and the shell is internal in the cuttlefish. It is worth mentioning here that the cephalopods have one of the most highly developed eyes of all invertebrates and certainly the most highly developed brain. Like most other mollusks, cephalopods have gills in the mantle cavity, but unlike other mollusks, they also have a well developed heart and a closed circulatory system, which support their very un-mollusk-like high metabolic rate. While the squids, cuttlefish and nautiluses are active swimmers, octopuses have taken up a more sedentary mode of life and usually lie near the entrances of their dens in rock crevices,

waiting for prey. Like octopuses, cuttlefish are capable of ejecting "ink" when threatened.

Like the squids and octopuses, the cuttlefish also is capable of changing color according to mood and need. In fact, it is believed that these animals utilize color changes for communication, to some degree, at least. Certain tropical species have the advantage of extra large chromatophores (pigment cells) that provide a range of color changes from different shades of red to absolute black. *Loliguncula brevis* is one such species, and it is imported occasionally from the Gulf of Mexico.

Since the cuttlefish colors change with mood and background, the hobbyist

A magnificent photograph of the cuttlefish *Sepia latimanus*. Photo by Bruce Carlson.

won't be selecting his species on the basis of color. A cuttlefish is the type of animal that you keep for its personality and because it is biologically interesting. It is different from the more familiar octopus in several ways. It has ten tentacles, although it is difficult to see all ten in a live animal, as the two that specialize in capturing prey are completely retracted, and the others are usually kept somewhat retracted. It is these short-looking tentacles that give the cuttlefish the "old man" appearance, as they give the impression of a beard.

The breeding of a cuttlefish is accomplished in the same way as for other cephalopods: a sperm packet is placed under the female's mantle with a specialized tentacle. However, I have not heard of any specimens reproducing in captivity. Obviously, hobbyists have their hands full just keeping one specimen, as the most commonly available species attain nearly a foot in length. (There are many small tropical species, though, that attain a size of barely six inches.) I am not sure how well individuals tolerate each other, but I have seen a tank with three specimens in a public aquarium. Very likely, the captive breeding of any cuttlefish species is in the distant future, as we are just now entering the stage, with all of our modern equipment, in which we can keep such marvelous animals and do so with good success.

Cuttlefish may look cuddly and resemble somewhat a little old man with a beard, but they are efficient

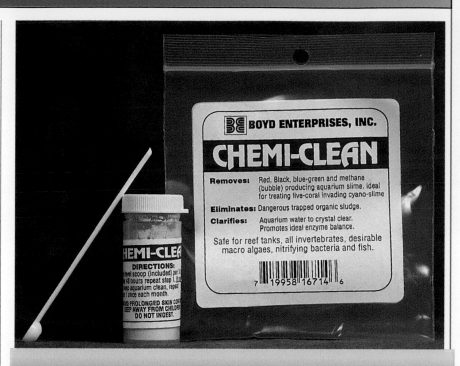

Products for eliminating slime-producing algae and organic sludge while cleaning the water are available at shops selling marine aquarium supplies. Photo courtesy of Boyd enterprises.

predators and among the most fascinating of aquarium exhibits. And you can tell your more naive and "non-combatants" in the marine world that you have captured an alien being from its flying saucer!

NON-PHOTOSYNTHETIC SESSILE INVERTEBRATES

Although reef tank hobbyists have had good accomplishments with photosynthetic anemones, corals and clams by using sun-like lighting and filtration that provides super-clean water, their success with certain corals and sponges has been less impressive. One reason for this is that many of the most beautiful soft corals and sponges tend to grow in the deeper waters of the tropics. They are not dependent upon

photosynthetic zooxanthellae; in fact, they lose out in competition with the animals that are. They don't grow quite as fast, and they are sensitive to the light. One way around this is to place such organisms in a shady area of a regular reef tank. But with its powerful lighting, sometimes shady areas can be difficult to find.

For such reasons, a different type of reef tank is worth considering. Everything will be nearly the same. The main difference is that you will need only enough lighting to view your specimens. You will not have to worry about high intensity or actinic lighting, but you will still need your trickle or fluidized bed filter and protein skimmer. And you will also want to make use of live rock. Such a tank is worth setting

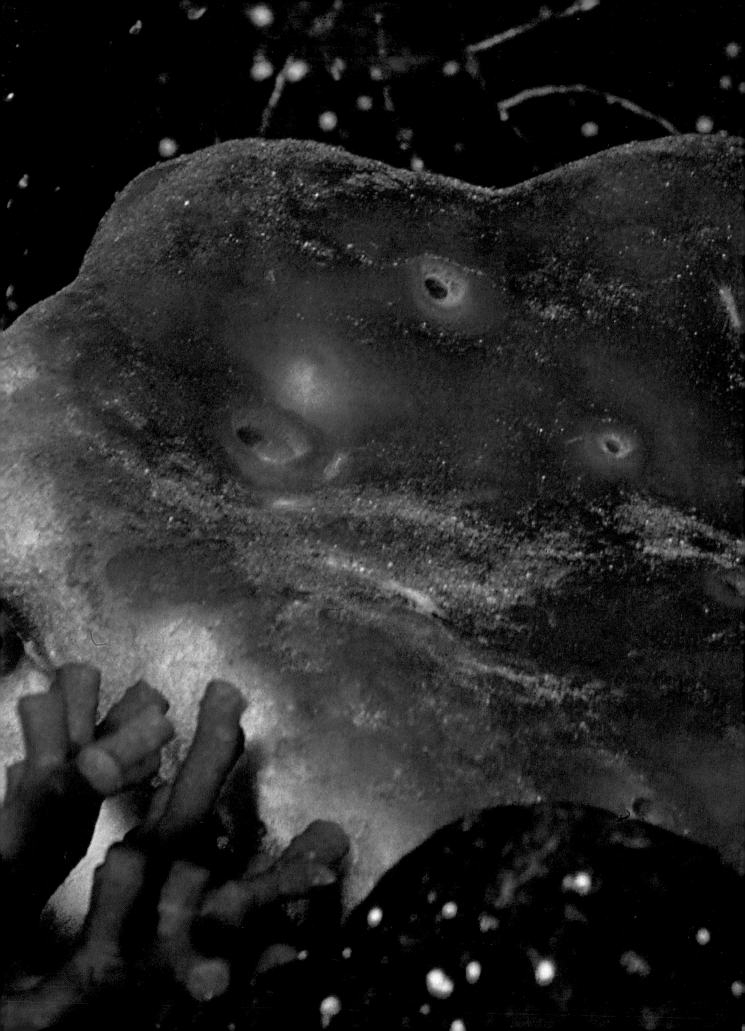

Petrosia ficiformis, an encrusting sponge. Photo by MP&C Piednoir.

up, for some of the deepwater corals and other invertebrates are among the most colorful. I well recall early divers utilizing lights at depths of a hundred feet or more and commenting on the colors that their bright lights brought out in these marvelous creatures found at such depths. The question was always "Why is something so colorful living where it can't be seen?"

As I said, everything will be very similar in this tank except for the lighting. One other difference is in the feeding of the corals. The photosynthetic corals often do not need feeding, as they obtain sufficient nutrition from the photosynthetic zooxanthellae in their tissues. Since the deepwater corals don't have zooxanthellae, they strain the water at night for plankton, and it is not just supplementary nutrition for them. (It should be pointed out that not all of these corals are necessarily from deep water; some simply grow in shady areas or in caves.)

The best way to feed these corals is with a turkey baster. Simply purchase a squeezebulb baster from the housewares section of a department store. Thaw the food in a plastic cup and draw it into the baster. Place the baster in the aquarium and direct a small puff of food at the coral. Do so gently, as too strong a blast may cause the corals to withdraw in self protection, as during a storm. It is also a good idea to feed at night, as that is more natural for them. However, you will notice that when you put food in for your other animals, the corals will often open their polyps from sensing the food in the water, even in the daytime.

The deepwater reef tank will be set up in a very similar fashion as the standard reef tank. One of the joys of a deepwater reef tank is a relief from concern

Pet shops sell chunks of living reefs which sprout into magnificent spectacles of life when placed into a proper marine environment. These rocks are called *live rocks* in the aquarium trade. Photo by U. Erich Friese.

The Jamaican sponge *Haliclona rubens*. Photo by Dr. Patrick Colin.

about problem algae, since it is possible to simply deprive them of light. Still, we want good water quality and low phosphate levels, and that is why we still utilize a protein skimmer and good biological filtration. However, we won't have to be so concerned about limiting the feeding of our animals. That is indeed fortunate, since we will be feeding them more—and more often.

Just what will we keep in this deepwater tank? Things aren't going to look all that much different except that we will have many corals that will have our fellow reef hobbyists scratching their heads in puzzlement and bulging out their eyes in

amazement at all the colors. And there will be an unusual predominance of other types of animals. Sponges, for example, are a large group of invertebrates that thrive in a deepwater reef aquarium. Some examples are the blue crust sponge (*Haliclona permollis*), the yellow cup sponge (*Verongula* species), the violet barrel sponge (*Siphonochalina* species), the yellow tree sponge (*Raspailia hispida*), the moon sponge (*Cliona* species), the purple sponge (*Lotrochata purpurea*), and the horse sponge (*Hippospongia communis*), to name just a few of the most beautiful species.

Sponges are filter feeders that feed on the smallest

particles of food. They die if they become coated with algae or clogged with particles. There will be no need to try to feed the sponges directly. They will get sufficient nutrition from the residual amounts that circulate in the tank and from protozoans that naturally occur there.

If handled properly, sponges will grow and multiply throughout your aquarium. If you decide to present the excess to a friend or transfer them to another tank, be sure to bag them under water, as they are extremely sensitive to being exposed to air.

Some of the most beautiful invertebrates are the soft

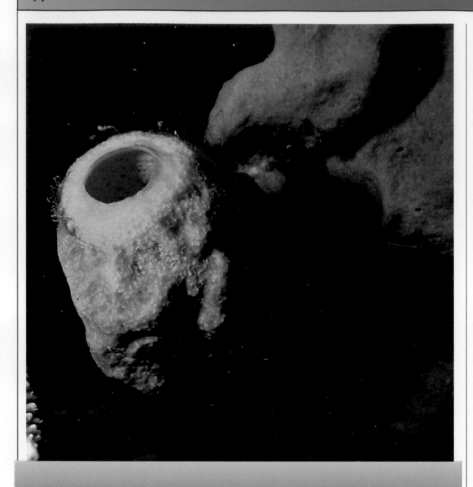

Above: *Verongia fistularis,* a Caribbean sponge. Photo by Dr. Patrick Colin. Below: *Haliclona permollis,* the purple breadcrumb sponge. Photo by Dr. Patrick Colin.

corals. Good candidates for the deepwater reef aquarium are the yellow *Dendronephythya aurea,* the red and white *Dendronephthya divaricata,* the orange-red *Carotaleyon sagamianum* and the pink and white *Scleronephthya* species. Some soft corals are fully expanded during the day, but most are nocturnal. Soft corals are easy to injure, and they must be placed in the aquarium with great care to avoid damaging them. For that reason, it is a good idea to purchase specimens that are still attached to their rock base.

Gorgonians are some of the most common cnidarians on the reef, and they depend on the currents to bring them food and to keep them clean. Hence water movement is just as important, if not more so, as it is in the standard reef tank. Alternating powerheads or a good strong laminar flow of water through the tank are particular assets. A beautiful sight is a red finger gorgonian (*Corallium rubrum*) with its lacy white polyps waving in the currents of an aquarium. As beautiful as they are, it figures that gorgonians would require the best water quality, and they are the earliest indicators of trouble in your aquarium. A simple glance at the gorgonians will let you know whether the pH has dropped or the protein skimmer needs cleaning. Look on the bright side, though. You now have an early warning system!

An important point here is that all of these gorgonians I have mentioned are commonly available in the reef hobby, but it is not always emphasized that these delicate animals are more likely to prosper in a tank without the high-intensity lighting. For that reason, many of your fellow reef hobbyists will be wondering what magical touch you have that allows you to keep such "delicate" animals. One of the things that they will want to check is your lighting, and imagine their surprise when they discover you have

and another is that most of the deepwater corals tolerate nutrient-rich water somewhat better, although it must be emphasized, once again, that water quality has to be maintained just as in the photosynthetic tank. Some of the fish that can be kept are royal grammas, combtail blennies, coral gobies, neon gobies, mandarin dragonets, sailfin gobies, jawfish, *Anthias* species, neon

lighting that is the mainstay of the standard reef tank. The same steps will be taken in setting up the tank. Just eliminate the step of turning on the high intensity lighting, and you don't have to worry about having your lighting on a timer. One of the secrets to the deepwater tank is keeping the lighting to short durations, so you most assuredly want your tank in a dark place in your home (that

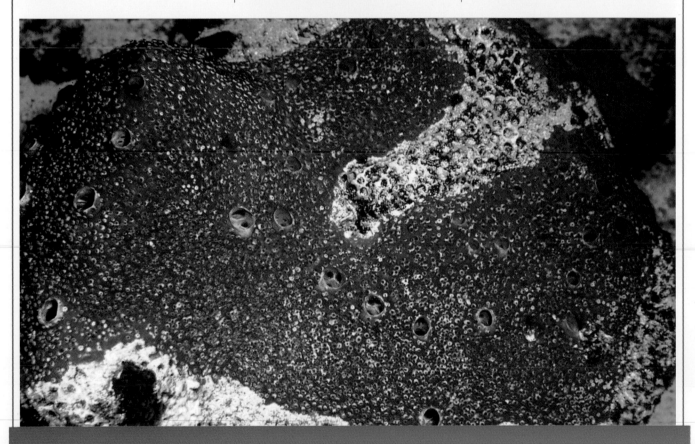

The Puerto Rican sponge *Cliona delitrix*. Photo by Dr. Patrick Colin.

"ordinary" aquarium lights!

Naturally, you will want fish in your deepwater aquarium, too, and the fact is that you can keep more of them in such a tank. One reason is that the temperature is slightly lower (between 70 and 75 degrees),

wrasses, leopard wrasses and lined wrasses, to name just a few.

Setting up a deepwater reef tank is going to be very similar to setting up the standard type. It is just that we don't have to worry about the sunlike high-intensity

is, at the very least, you don't want it in a sunny area!) so that you can keep more control over the lighting.

The reason limited lighting is important is that algae are a real enemy to sponges and soft corals. The algae can block the pores of the

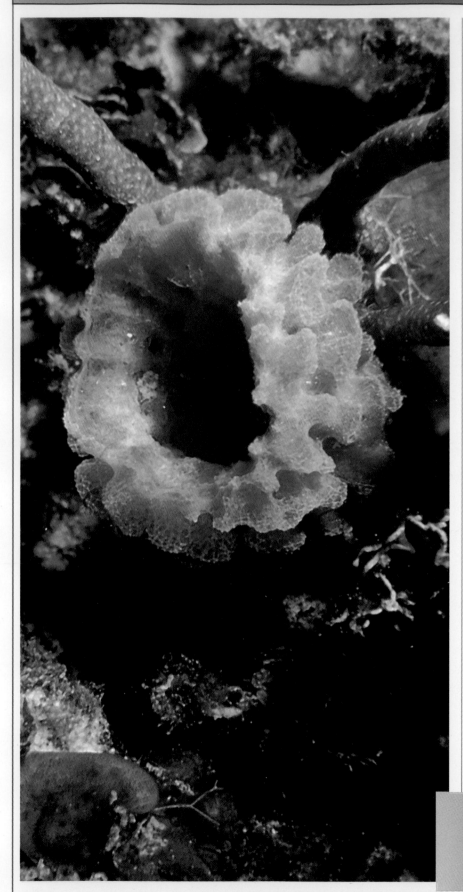

The sponge *Callyspongia plictifera*. Photo by Dr. Patrick Colin.

sponges so that they simply starve to death. Soft corals defend against algae by secreting a slime that lets them simply slough off an entire outer coat, thus ridding themselves of encrusting algae, but they should still be protected.

To be perfectly clear, algae can be a problem in a "common" reef tank, too, where they have to be controlled by strictly curtailing phosphate levels, and this can often be a challenge. That is because you need the bright lighting for the corals, and you can't turn it off. In the case of the deepwater tank, however, algae can be easily controlled simply by starving them to death from lack of light. Blue-green algae (cyanobacteria) can be a problem even in limited light, and that is one reason for maintaining high water quality; the high water quality will discourage the growth of such organisms.

Admittedly, such tanks haven't become the craze that the standard reef tank has, but that very fact helps make them special. Certainly this type of tank can hold its own in beauty to any other tank. But it also challenges any other tank in the interesting category, too.

ARTHROPODS

The class of insects is one of the reasons that the phylum Arthropoda is so large, as insects are the largest class in the animal

Closeup of the soft coral *Dendronephthya*. Photo by Takemura and Suzuki.

kingdom in terms of the number of species contained. Insects are terrestrial, and one reason for their success is their power of flight and also because they were one of the first animals to colonize the land. In the sea, the arthropods are also well represented; however, most are larger than most insects. In fact, lobsters look like "giant bugs" to youngsters who haven't seen them before (and even to young damsels, to judge from their actions upon seeing one). There are also giant crabs in the ocean. But one of our objectives with marine aquaria is to attempt to find specimens that stay a reasonable size.

Although lobsters and crabs superficially resemble insects, they can be distinguished by the fact that insects have three body parts and six legs. Animals such as lobsters have ten legs. Most of the arthropods in the ocean are crustaceans. The following are some that are popular in the aquarium.

CRABS, LOBSTERS AND SHRIMP

The class Crustacea includes some of the most charming and beguiling of reef inhabitants. It also includes animals, such as the mantis shrimp, that have no place in the living reef tank, but somehow often end up there—much to the exasperation of the hobbyist! Often, though, the charm and beauty of a reef tank is topped off by a display of how a favorite crab will take food from its keeper's fingers. It might be of interest to the "reef tankers" to know of some of the crustaceans that are available to them. It might be of even greater interest to know how the different species adapt to a reef tank.

ARROW CRAB (*Stenorhynchus seticornis*): This species is not being listed first simply because of alphabetical order. It has much to recommend it, not the least of which is the fact that it will eat bristleworms! It gets its common name from the distinctly triangular, arrowhead-shaped body. This feature, together with its very long and thin legs, results in an impression of a giant spider. It gets to be about six inches, counting the long legs, but it is not considered a threat to corals, as its dietary specialty is burrowing worms.

Hobbyists with featherduster worms in the tank may want to exclude this species, as it will go after them, too. All in all, this is a desirable species in that it is easy to maintain and has a pleasing "personality." One of the few drawbacks to the arrow crab is that it is a little large for some tanks. The species is quite territorial in regard to others of its species, so only one should be kept to a tank.

ANEMONE CRAB (*Neopetrolisthes ohshimai*): This Indo-Pacific crab is one of a small group of porcelain crabs that have evolved an immunity to anemone stings. Like clownfish, anemone crabs can live among the tentacles of various large anemones and receive protection from predators. They measure barely an inch across the carapace, and they are surely among the best of the crabs as potential reef candidates.

The crabs live in the same types of anemones as the clownfish and will use their well developed claws on any clownfish that tries to evict them; however, the clownfish are not as aggressive toward them as they are toward

crabs that tend to feed upon the anemone. The anemone crab feeds upon small particulate matter, including plankton, and it may help keep the anemone clean by scavenging debris from among the tentacles. The crab has small feathery projections on its jaw processes to process particulate matter. As is the case with most other crustaceans, the anemone crab is particularly vulnerable to predation right after it has shed its carapace; hence, hiding places are needed if you aren't keeping it with a suitable anemone.

Several other species of this genus also are suitable for the reef and other invertebrate aquaria.

BOXING CRAB (*Lybia*

The arrow crab, *Stenorhynchus seticornis*. This is a valuable animal in Nature where it consumes bristle worms. Its main diet is burrowing worms. Photo by U. Erich Friese.

A Pacific anemone crab, *Neopetrolisthes ohshimai*. Photo by Mella Panzella.

tessellata): This little crab has to be in the running for being the most charming—and most bizarre—of all animals. This crab reaches barely an inch in length. Still, he is not a fellow to be trifled with. Boxing crabs collect a tiny anemone in each claw and actively wave them at encroaching predators as a warning (the boxing effect). So we see that the boxing crab is not only a tool-using animal but also a user of most unusual tools: living anemones. When the crab sheds its exoskeleton, it carefully puts down the anemones, setting them aside until the new shell hardens. Then the anemones are picked up once again and pressed into service.

The boxing crab is not fussy as to food and is one of the species that will take food from the fingers of its keeper. There are many suitable species of boxing crabs that make good reef residents, and nearly all of them are in the genus *Lybia*. And all of them are more than just comical; they are fascinating to observe and have a fascinating life history as well.

RED DWARF LOBSTER (*Enoplometopus occidentalis*): Most of us are primarily familiar with the lobster species we use for food, and, of course, those are much too big for a reef aquarium. This species only reaches a size of about four inches, but it is still a questionable candidate for the reef tank. It is a good scavenger, but it is also predatory on crabs and may seize fish at night when they are sleeping. It will defend its territory against other species of lobster, so this truly is a questionable species in spite of its charm. Nevertheless, there are situations in which it could be a suitable aquarium inhabitant. For example, a hobbyist keeping a tank of large anemones, clownfish and some of the anemone crabs would not have to worry about having a red dwarf lobster harming any of them.

PURPLE SPINY LOBSTER

The boxing crab *Lybia tessellata* is a real oddity for the mini-reef aquarium. It collects an anemone in each claw and threatens predators with them as you can see in this photo! Photo by Scott Johnson.

The red dwarf lobster, *Enoplometopus occidentalis*, is sometimes tolerated as a possible mini-reef invertebrate because of its relatively small size and bright color. It is too aggressive against crabs. Photo by Alex Kerstitch.

(*Panulirus versicolor*): Another lobster species that is a possible candidate is the purple spiny lobster, *Panulirus versicolor*, which does not have the large claws. It reaches a size of about eight inches and is not as predatory as the red dwarf lobster, but it can do accidental damage just because of its size.

All lobsters, incidentally, seem to be easily startled, and, when they are, they shoot backwards with a quick flip of the tail and other appendages. A maneuver like

that can cause damage to some of the more delicate corals, so it is rare to see a lobster in a reef tank. The hobbyist who wants to keep any of the lobster species must take into account that most of them are going to hide during the day anyway since they are nocturnal (although, to be honest, most can be coaxed out with food). And once you put a specimen in and it creates any sort of problem, you are going to have a devil of a time getting it out. Plan on a general dismantling of the tank in

order to remove a lobster.

BANDED CORAL SHRIMP (*Stenopus hispidus*): Here is a species that looks like a lobster and has all the advantages of a lobster, but without having any of the drawbacks. First, it only attains a size of about two inches. It stays out in the open in the daytime (once it has become accustomed to the tank), and it is compatible with other species of shrimp. It may fight with other members of its own species, but it is possible to get a compatible pair. It has

Above: The purple spiny lobster, *Panulirus versicolor.* This beauty is not predatory, but it grows twice as large as the red dwarf lobster. Below: The banded coral shrimp, *Stenopus hispidus,* looks like a lobster and acts like a lobster but is, in reality, a shrimp. It is very colorful and not aggressive. It is highly valued for the mini-reef tank. Photo by U. Erich Friese.

personality and will gesticulate and wave for attention with its claws. It is a hearty eater and a good scavenger. As is the case with most other crustaceans, it is quite vulnerable after it has shed its exoskeleton and thus should have plenty of hiding places into which it can retreat while its carapace hardens.

CLEANER SHRIMP (*Lysmata amboinensis*): Not only is this a good looking species, but it is particularly interesting because it is involved with cleaning symbiosis, almost a trademark of the ocean. The common name for these very attractive and sociable shrimps comes from their natural cleaning behavior. In the wild, on Indo-Pacific reefs, they will pick parasites from many species of fish, including large groupers and moray eels.

This species is long lived in the aquarium. A group of five or so can be kept amicably together, and they will even spawn in the reef or invertebrate aquarium. The females develop large quantities of green eggs under the abdomen. Raising the pelagic young is another matter. It could probably be done in a tank kept just for that purpose, but in the reef tank the tiny pelagic young are simply part of the menu of the filter feeders.

CANDY SHRIMP (*Rhynchocinetes uritai*): Here is another attractive shrimp that is best kept in a group of about five or so. Single specimens tend to hide most of the time. This shrimp only reaches a length of about an inch and is quite defenseless

The cleaner shrimp *Lysmata amboinensis* is a wonderful animal which cleans parasites from many large fishes like groupers and moray eels. Rarely (but occasionally) these larger fishes eat their cleaners. Photo by U. Erich Friese.

against other crustaceans. This species is quite hardy, easy to keep and generally trustworthy with the corals and other reef animals. Those traits, combined with the candy shrimp's impressive appearance, make them excellent candidates for the reef tank.

HARLEQUIN SHRIMP

(*Hymenocera picta*): This species is often offered for sale and is nearly irresistible, as it is gorgeous. In fact, they are sometimes sold as pairs by dealers. As such, they certainly make a nice display. Unfortunately, they are very specialized feeders, dining solely on starfish. Hence, unless you have a supply of starfish for feeding purposes, your harlequins will simply starve to death. In view of the fact that we have occasional plagues of crown of thorns starfish that decimate coral areas, it would seem that those starfish could be collected and frozen for use as food. While I am not in favor of decimating any species, surely overpopulated species that are doing harm could be used.

ANEMONE SHRIMP (*Periclimenes imperator*): There are several species of anemone shrimps, but most of them are transparent or of a color that makes them difficult to spot. They live among the protective tentacles of anemones, much like clownfish and anemone crabs. This species hails from the Red Sea and Indian Ocean; it is most often crimson with a white pattern over the top, although coloration does vary from individual to individual. It reaches a length of only an inch and should not be kept in the same anemone with other crustaceans, such as the anemone crab.

Other Crustaceans

BARNACLES: Occasionally barnacles of the genus *Lepas* are offered in aquarium shops, but they sometimes have them listed as mollusks, as that is just what gooseneck barnacles appear to be. But what look like shells of bivalves are actually an armored development of the chitin exoskeleton that is common among arthropods, including crustaceans. The barnacle's feet are modified with feathery appendages so that they can scoop food into the mouth.

Harlequin shrimp, *Hymenocera picta*, are lovely additions to the marine aquarium, but their sole diet is starfish. Photo by U. Erich Friese.

An anemone shrimp at work. *Periclimenes imperator* is one of the three known species of anemone shrimps. The other species are transparent and thus not attractive for aquarists. Photo by Dr. Patrick Colin.

Goose barnacles, *Lepas*, though rarely available, make excellent additions to the invertebrate mini-reef exhibit. Photo by Aaron Norman.

MANTIS SHRIMP (*Odontodactylus* species): Many of these species are quite colorful, and they certainly do have personality. But they're listed here more as a warning than anything else. They are a reef keeper's worst nightmare. They are fierce predators, taking fish, corals and just about anything else you have in your tank. To top it off, they are known unaffectionately as "thumbsplitters" by fishermen. That nickname tells you that

certainly tunnel through coral.) Although mantis shrimp have no place in a reef tank, they are of sufficient interest to make a good display in a tank of their own.

In spite of the fact that I have concluded my sample of crustaceans with a horror for the reefkeeper, there are countless other crustaceans that are beautiful or unbelievably ugly. (Even most mantis shrimp are of quite attractive coloration—which demonstrates that your

Not only does the armored "shell" of the barnacle protect it from enemies, but it also keeps it from drying out when it is exposed to the air if the barnacle is above the water line when the tide goes out.

Obviously, barnacles must be fed the same type of planktonic food that is offered corals and other filter feeders. The secret to success is to feed small puffs of the thawed-out food from a meat baster several times a day. (Or evening, if your barnacle feeds only at night; however, most barnacles will begin feeding when food is in the water.)

Most hobbyists are only going to get barnacles accidentally as passengers on live rock, and it is difficult to give them frequent enough feedings to have them survive more than a few months. It is ironic that animals that are such a pest to boat owners and so common among the docks and rocks of the seashore should be so challenging to keep alive. The secret is in frequent feedings of planktonic food.

The mantis shrimp, *Odontodactylus*, should NOT be kept in a mini-reef tank. They are killers of almost every living thing to be found in a mini-reef tank including fishes and corals. Photo by Kok-Hang Choo.

they can even do damage to the hobbyist. And just when you think things couldn't be worse, they are! If you try to catch a specimen that has once got into your tank, you will find yourself chasing a veritable will o' the wisp, as these things can seemingly tunnel right through solid rock. (Actually, that is an exaggeration, but they can

mother was right when she told you not to be beguiled by beauty!) The fact is that the crustaceans take the prize in all directions. They have fascinating behavior, and there are so many different species that this little sketch of a few random varieties barely scratches the surface of the fascinating world of crustaceans.

STAPLES AND ODDITIES

The emphasis in this book has been on invertebrates without regard to whether they are compatible in a reef tank. There has been an emphasis on non-photosynthetic invertebrates. That is why we covered sponges and some of the non-photosynthetic corals. Still, excluding corals, it would be of interest to review some of the popular invertebrates in the reef tank, and some of the oddities, too, as they don't have to be restricted to just a reef tank. Blue-legged and red-legged hermit crabs, for example, are much prized in a reef tank, for they don't threaten any of the animals, and they do help control the algae problem that can be the bane of reef hobbyists.

The *Tridacna* clams are probably the largest clams in the world. They are found on most South Seas reefs, represented by the species *Tridacna gigas*. Small specimens are ideal for the mini-reef tank. Photo by Gunter Spies.

TRIDACNA CLAMS

Almost all reef tanks support clams of the genus *Tridacna*, and with good reason. Clams such as *Tridacna crocea* have beautiful mantles and are usually about four inches long, although they are capable of reaching a foot in length. Fortunately, these clams are quite slow growing, and they thrive in the reef aquarium. Like all other clams, they are filter feeders; however, they are able to thrive even without food if they have bright actinic lighting available for the zooxanthellae in the tissues of their mantles.

Tridacna gigas is the South Seas giant clam that is so often depicted by artists with a diver's foot clamped in its maw. This is an exaggeration, as skin divers investigate these animals without danger to themselves. Although all *Tridacna* clams are capable of trapping a small fish or invertebrate inside their shells, they don't snap shut like a mousetrap, so a diver would have plenty of time to remove a foot. In any case, most of these clams get their nutrition from the zooxanthellae in the mantle, so they are of little danger to any of the other inhabitants, and they look great in the aquarium.

Surprisingly, the giant clam, *Tridacna gigas*, is kept by many reef tank hobbyists, but the hobbyist who has one had better be ready to eat it once it attains its full girth, as it can weigh several hundred pounds. One of the reasons that these clams are kept is that they grow slowly, and when they are "only" a couple of feet across they provide an excellent centerpiece for the aquarium.

OTHER BIVALVES

FLAME SCALLOP (*Lima scabra*): The so-called flame scallop hails from the Caribbean Sea, and it is popular because of the coloration of the tentacles around the shell. These tentacles are usually a bright orange, and when the scallop propels itself by jet

Above: The flame scallop, *Lima scabra*, from the Caribbean Sea is an aquarium favorite. Photo by Courtney Platt in the Cayman Islands. Below: The Atlantic thorny oyster, *Spondylus americanus*, is one of the best-camouflaged mollusks on the reef. Photo by Dr. Patrick Colin.

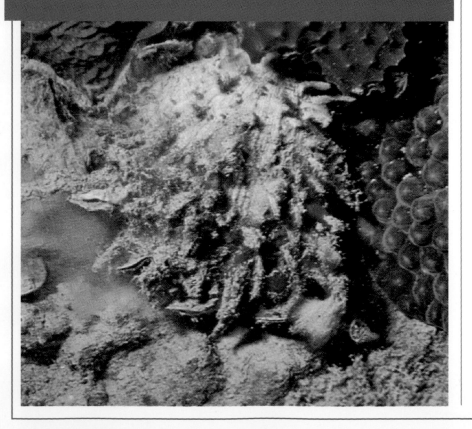

propulsion (from the force of the water expelled when the shells are closed repeatedly, like a bellows), the fringes on the mantle look very much like an Indian headdress, flowing in the "wind."

Although these bivalves are commonly available, they need frequent feedings of plankton in order to survive long term. This should be done several times a day with a baster. Frozen plankton can be utilized. Simply thaw the plankton in a paper cup and utilize the baster for transporting it to the clam. Unlike the *Tridacna* species, the flame scallop does not have photosynthetic zooxanthellae and therefore needs the frequent feedings of plankton.

THORNY OYSTER (*Spondylus americanus*): This is another bivalve from the Caribbean. Both shells, or "valves," of this species have long, thornlike extensions; when cleaned, the shell is attractive and a favorite with shell collectors. Collected from the ocean, the thorns are often heavily covered with growths of sponge, hydroids and sea squirts.

Like the flame scallop, the thorny oyster needs frequent feedings of planktonic food in order to do well in the aquarium, and it does not obtain any nutrition from photosynthetic organisms.

TUBE WORMS

Tube worms are variously known as "featherduster worms" and "fan worms." They are annelids (phylum Annelida) of the family Sabellidae. Each tube worm's body is encased in a parchment tube buried in the

The reason that some worms are called *featherdusters* is the feathery appendages of some of the Sabellidae worms. Photo by MP&C Piednoir. Below: The tubeworm or featherduster worm *Sabellastarte magnifica*. Photo by Mike Mesgleski, Cayman Islands.

A magnificent Christmas tree worm from the Caribbean, of the family Serpulidae. Photo by Courtney Platt in the Cayman Islands.

substrate, with the feathery head extended for feeding. At the approach of danger, the feathery tentacles are rapidly withdrawn into the tube. Tube worms are fairly hardy animals, although they must be fed daily (or more often if possible) on the plankton foods that are provided for sale, usually in a frozen form.

Tube worms will occasionally reproduce in the home aquarium. Spawning consists of emitting the eggs and sperm, which are so small that they look much like "smoke." The feathery appendages are shed after the spawning, but they eventually grow back.

CHRISTMAS TREE WORMS

Christmas tree worms get their name from the fact that the symmetry of their feathery appendages is such that they resemble a Christmas tree. The fact that these appendages are of different coloration may also have contributed to the name. Christmas tree worms are usually sold as a special live rock, normally called Christmas tree rock. In any case, they are annelids (family Serpulidae); they differ from the tube worms discussed above in being generally smaller and in having the tube usually embedded in coral or coral rock. They require the same care as tube worms, with occasional feedings of planktonic food. A main concern with both is to avoid

Christmas tree worms of the genus *Spirobranchus*. Photo by Dr. Gerald Allen.

predators, either fish or invertebrate. Certain sea hares and other gastropods are quite predatory on these worms. Whenever you purchase any gastropods or fish, for that matter, check with your dealer to make sure that the new additions are not predatory on Christmas tree worms.

TURBO SNAILS

There are a number of herbivorous gastropod snails of the genera *Astraea, Turbo, Trochus, Nerita, Cerithium* and *Calliostoma* that are referred to as "turbo" snails. A point of fact is that the actual snails of the genus *Turbo* are not as preferred as some of the others, for they grow too large, up to three inches in length, and they can knock over some of the corals in their food-seeking activities. For that reason, snails that are really of the genus *Turbo* may be preferentially placed in non-reef invertebrate aquaria as a means of algae control and because of the interest in the animals themselves.

More desirable species for the living reef tank are snails of the genus *Astraea* of the Caribbean. They are long lived and don't grow too large to be a problem to the corals. For algae control, about one to a gallon is recommended for smaller aquaria, with proportionately fewer being needed in larger setups, because the amount of rockwork exposed to bright light doesn't increase at the same rate as water volume.

OTHER GASTROPODS

QUEEN CONCH: The queen conch, *Strombus gigas*,

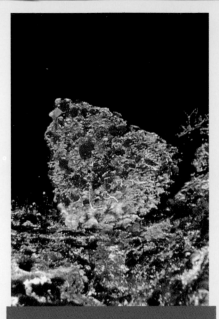

Above: *Astraea tecta*, the American star-shell, is a gastropod mollusk from the Caribbean. This Puerto Rican specimen is coated with algae. Photo by Dr. Patrick Colin.

Above: A turbo snail, *Calliostoma* species, from the Cayman Islands. Photo by Courtney Platt.

would be an ideal univalve for the reef tank if it were not for the large size that they eventually reach. Still, they

are hardy and travel all over the tank taking care of all types of algae, including even blue-green (cynobacteria) algae. Since queen conchs get so large and have such a strong foot, they are not popular in tanks containing corals, but they are great for

A very algae-laden mini-reef featuring blue starfish, *Linckia*, and *Amphiprion* clownfish. Photo by Dr. Cliff W. Emmens.

The Hawaiian conch *Strombus hawaiiensis*. Photo by Scott Johnson.

compressed and quite small crustaceans that stay hidden during the day. They live among the algae and are generally omnivorous, but some are strictly herbivorous. The way to see them is to use a flashlight at night when the lights are out, and you can identify them by their arched backs. One reason that certain fish, such as the mandarin fish, can be kept alive in the reef tank even though they do very poorly in a fish-only tank is that they are specialized feeders, subsisting primarily on amphipods. If the tank is large enough and there are enough live rocks, the supply of amphipods never runs out, and the mandarin fish prospers. And your friends who couldn't keep them think you are a genius!

SEA SQUIRTS

Most hobbyists only see their closest living invertebrate relative on live rock that they have purchased. Occasionally, however, large specimens are offered for sale by aquarium shops. Some of them are

the invertebrate tank. Not many other animals can hurt it, and it won't bother anything else—at least, not on purpose! I kept a small one (only an inch long), and it had not attained much size in two years when I let it go to someone else. Others may have different experiences, as I made no effort to feed it, because it seemed to find enough on its own to keep it in apparently good shape.

SEA HARES

The sea hares get too large for the reef aquarium and, like snails of the genus *Turbo*, they will tip over coral heads. They sometimes "fly" through the water, using undulations of the body to propel themselves, and they can come down on coral heads. They are fascinating in non-coral invertebrate tanks, however, and they subsist primarily on algae.

Unfortunately, some species also will eat your featherduster worms.

The sea hare gets its name from being about the size of a rabbit when fully grown and from having little projections that look like rabbit ears. Of course, it is a mollusk without a shell. A good candidate for the aquarium is the common Caribbean sea hare, *Aplysia dactylomela*.

AMPHIPODS

Amphipods are introduced accidentally by the addition of live rock. They are laterally

The sea hare *Aplysia californica* from the Pacific is closely related to the Caribbean *Aplysia dactylomela*. Photo by Mark Smith.

quite colorful, although the most usual color is something resembling transparent cellophane. The problem is trying to keep them alive with frequent feedings of planktonic food. I've done it with temperate specimens, so I am assuming that it can certainly be done with tropical species. Not many people have been motivated to give them what they require by way of food, because sea squirts don't look much different from sponges, even though there is a world of difference. Remember, sea squirts have a hollow stiff notochord in their larval stage. As adults, they actually have organs and muscles, things sponges don't have.

Some believe that vertebrates may have developed from sea squirts (or a sea squirt-like ancestor) through a process called neoteny. That means that the youthful (in this case, larval) form is retained in the course of evolution. This is why I called the sea squirt our closest relative among the invertebrates—even though certain crabs may look more like many of us!

STRANGE BEDFELLOWS

It is typical of ocean creatures for strange partnerships to evolve. One of the best known is that of the anemone and the clownfish, but there are many others in the ocean. One that can be displayed in the aquarium is the strange partnership between burrowing pistol shrimp and watchman gobies (sometimes called shrimp gobies). The association works like this. The shrimp is an adroit digger, and it carves

The mandarin fish *Synchiropus splendidus* is popular in the reef aquarium because it eats amphipods. Photo by Dr. Herbert R. Axelrod.

out a labrynth of passageways beneath the surface of the substrate. It is constantly repairing the structure, pushing out gravel and keeping the entrance open. It feeds upon detritus and other organic matter that finds its way into the cave. The goby gets a home out of the deal and shelter from predators. For a long time, observers didn't understand what the shrimp received from the association. Then it was observed that the shrimp keeps in contact nearly constantly with the goby by maintaining tactile touch with one of its antennae. Thus, the goby acts as something of a seeing eye dog for the shrimp, which is nearly blind. When danger threatens, the goby flicks its tail, just once if the danger is not too great. Several flicks mean that the shrimp had better retreat deep into the tunnel, and the

goby follows right after it.

Many reef hobbyists don't like to keep gravel and sand in the tank, so it is difficult for them to display this shrimp and fish relationship. One way around that problem is to provide PVC piping with fittings. Of course, this doesn't look too natural on the bare bottom, but it could

The goby *Cryptocentrus koumansi* and an alphaeid shrimp. Photo by Dr. Gerald R. Allen.

A wonderfully colored nudibranch, *Chromodoris norrisi*. Photo by Alex Kerstitch

be placed beneath the live rock.

If you don't have a reef tank and also don't have the type of gravel mix that you feel will be good for tunnel building, you can still utilize the PVC set up. Just bury it under the sand with the openings visible. It may drive the shrimp crazy that it can't change the structure of its tunnel, but it is doubtful that it will abandon it.

The effect of the display is that the shrimp is constantly pushing out sand, and the goby is hovering, "on watch," just outside the entrance. In nature, the goby feeds upon small crustaceans, such as copepods, and other items

that the current may bring it just outside the entrance. It is interesting to note that the shrimp ceases pushing sand at night. In the sea, that amounts to letting the entrance close. That means that the two are safe from prowling moray eels that come out at night and can fit down inside burrows.

Since the gobies are attracted to specific species of shrimp, it is best to buy your specimens as a "matched set."

COMMENTS

Again, we have barely touched the surface with the enormous number of invertebrates that can be kept

in the home aquarium. And we have not covered the filtration system that should be used with these animals. One reason for that is that the invertebrates will do well in nearly any filtration system. With the exception of the cephalopods, nearly all invertebrates have a low metabolic rate. For that reason, they don't put a big load on any filtration system, and they don't impact the quality of the water as much as so-called higher organisms do with their "speeded-up" living styles. In fact, if you keep only invertebrates or limit your fish to only two or three small specimens, your tank can be literally loaded with invertebrates. Who could ask for more?

Of course, some invertebrates are extremely easy to keep, while others are more challenging. That just means that it is difficult to duplicate natural conditions in the home aquarium. Often this is simply a matter of supplying the natural food. For example, we have not included nudibranchs, a very beautiful group of gastropods, because nearly all of them are very specialized feeders, feeding upon corals, tunicates or other nudibranchs, the types of things that are difficult to keep in constant supply in the aquarium.

We have tried to give a survey of invertebrates that you can properly care for and that will give you much joy, not to mention an education. Again, what more could we ask of our much-maligned invertebrate friends? Life with an invertebrate tank may at times be frustating—but it will never be dull!